ROBERTO ROSSI

INVENTORY ANALYTICS

OPEN BOOK PUBLISHERS

Author: Roberto Rossi Email: robros@gmail.com Website: https://gwr3n.github.io

Roberto Rossi, *Inventory Analytics*. Cambridge, UK: Open Book Publishers, 2021, https://doi.org/10.11647/OBP.0252

ISBN Paperback: 978-1-80064-176-1
ISBN Digital (PDF): 978-1-80064-177-8
DOI: 10.11647/OBP.0252

Dedicated to my family and friends.

ORÐ MÉR AF ORÐI, ORÐS LEITAÐI. VERK MÉR AF VERKI, VERKS LEITAÐI.

HÁVAMÁL, 140.

Contents

List of Figures

List of Tables

List of Listings

Preface

Inventory control is a thriving research area that plays a pivotal role, as a building block, in supply chain planning. For this reason, it attracts the attention of both industry and academia.

Selected topics from inventory control are regularly covered in academic programmes, at both undergraduate and graduate levels, offered by business schools, industrial engineering, and applied mathematics departments.

Problems faced by managers who engage with the challenges posed by inventory systems are generally simple to state, but complex to address. Obtaining good solutions to these problems requires a blend of expertise drawn from a variety of quantitative disciplines, such as operations research, economics, mathematics, and statistics.

The majority of existing books in inventory control theory adopt, in my view, an overly mathematical and abstract style of presentation. This style appeals to researchers in the area, but makes these books often inaccessible to practitioners, as well as to some business school researchers who have not received advanced mathematical training such as that offered by applied mathematics, computer science, or industrial engineering curricula. A book with a more applied, hands-on focus is missing.

This work aims to fill this void. It is aimed at those who want to learn the basics of modelling aspects of inventory control problems without needing to resort to the technical literature; at those who, despite lacking advanced mathematical training, want to access seminal findings in this field, and to apply well-established models by employing state-of-the-art solvers and modelling languages.

The book requires a working knowledge of Python; it is therefore aimed at readers who have, at the very least, taken a basic Python programming course. Apart from this, the book aims at stripping mathematical results to the bare minimum while preserving sufficient rigour, and at focusing on the practical relevance of these results in the context of the implementation of solution methods for problems typically faced by a manager who juggles with day-to-day inventory control challenges.

The book is structured as follows. It first provides a general introduction to inventory systems, followed by an overview of basic deterministic models. All these models are paired with their respective Python implementation, which can be tested on motivating examples that are presented throughout. After showcasing established models in deterministic inventory control, the reader is introduced to forecasting. Forecasting is often only briefly surveyed in existing books on inventory control; with the readers often directed to specialised textbooks, which are again often inaccessible to practitioners or individuals without suitable advanced mathematical training. However, forecasting is a crucial aspect of any practical inventory challenge. This work covers the most well-known forecasting models in a hands-on and visually appealing manner. The introduction of forecast errors paves the way to stochastic inventory control models, which are presented in the following sections. Once more, the most well-known stochastic inventory control policies are discussed in a hands-on fashion, with supporting code snippets and motivating examples. The last chapter briefly presents seminal results in the context of the control of multi-echelon inventory systems. Finally, an appendix provides the relevant formal backgrounds on a number of topics that are leveraged throughout the main chapters.

Introduction

This book originates as a collection of self-contained lectures. These lectures are divided into an introduction to inventory control, which outlines the foundations of inventory systems; followed by three chapters on deterministic inventory control, demand forecasting, and stochastic inventory control.

Beside Inventory, the title of the book refers to Analytics. This is nowadays a concept that has been inflated with a plethora of meanings, so that it becomes difficult to understand exactly what each of us means when we refer to it. The Cambridge Dictionary[1] defines Analytics as "a process in which a computer examines information using mathematical methods in order to find useful patterns." However, this appears to be quite a restrictive definition for our purposes.

[1] *Cambridge Dictionary*, `https://dictionary.cambridge.org/dictionary/english/analytics?q=Analytics`

To better understand the nature of Analytics, it is useful to observe that Analytics is often broken down into three parts: descriptive, predictive, and prescriptive. Descriptive Analytics is concerned with answering the question: "what happened?" Predictive Analytics is concerned with answering the question: "what will happen?" Prescriptive Analytics is concerned with answering the question: "how can we make it happen?" These are clearly complex questions that cannot be answered by mere *number crunching* on a computer: to answer these questions a decision maker must leverage soft as well as hard skills.

Fig. 1 This allegorical woodcut shows Rhetorica enthroned between Prudentia and Inventio; Willem Silvius, Antwerp, 1561 (Image by Anonymous, Wikimedia, public domain).

Many tend to think that the Analytics phenomenon is a recent development related to widespread availability of computing power. However, in his work "De Inventione," the Roman philosopher Cicero states that "there are three parts to Prudence: Memory, Intelligence, and Foresight." It is clear that Memory is the skill required to answer the question "what happened?"; Foresight, that required to answer the question "what will happen?"; and Intelligence, that required to answer the question "how can we make it happen?" It appears then that Analytics is just a contemporary rebranding of an art that has been known for millenia. *Prudentia* is the ability to govern and discipline oneself by the use of reason. *Inventio* is the central canon of rhetoric, a method devoted to systematic search for arguments. Incidentally, *inventio* also means inventory. In fact, when a new argument is found, it is *invented*, in the sense of "added to the inventory" of arguments. *Prudentia* and *Inventio* are the foundations upon which the art of Rhetoric stands (Fig. 1).

https://doi.org/10.11647/OBP.0252.06

It must not surprise us then that Analytics plays a prominent role in inventory management. Inventory management finds its roots into the practice of late medieval and early Renaissance merchants.[2] The invention of double-entry bookkeeping (alla Veneziana) is typically attributed to Frà Luca Pacioli (c. 1447 – 19 June 1517). Pacioli leveraged Johannes Gutenberg's new technology to disseminate and popularise accounting practices that had been in use among Venetian merchants for a long time. However, Pacioli did not simply disseminate existing practices, he reinterpreted these practices within the framework of Cicero's rethoric.[3] In "De Inventione," Cicero explains that there are five canons, or tenets, of Rhetoric: *Inventio* (invention), *Dispositio* (arrangement), *Elocutio* (style), *Memoria* (memory), and *Pronuntiatio* (delivery).

[2] Alfred Crosby. *The measure of reality: quantification and Western society, 1250-1600*. Cambridge Univ. Pr., 1997.

[3] Paolo Quattrone. Books to be practiced: Memory, the power of the visual, and the success of accounting. *Accounting, Organizations and Society*, 34(1):85–118, 2009.

Pacioli's "Tractatus de computis et scripturis" (1494, Fig. 2), is divided into two main sections: (i) the Inventory, and (ii) the Disposition — the influence of Cicero's work is apparent. Pacioli writes: "In order to conduct a business properly a person must: possess sufficient capital or credit, be a good accountant and bookkeeper, and possess a proper bookkeeping system." In "the Inventory," Pacioli writes "The merchant must prepare a list of his inventory. Items that are most valuable and easier to lose should be listed first. [...] The inventory should be carried out and completed in a single day. [...] The inventory is to include the day that the inventory was taken, the place, and the name of the owner."[4] In contemporary terms, Pacioli describes a so-called "physical inventory," the process by which a business physically reviews its entire inventory — as opposed to so-called "cycle counts," which focus on specific subsets of items. In "the Disposition," Pacioli describes the necessary books and rules to implement double-entry bookkeeping.[5]

Fig. 2 Dedication page of Pacioli's "Tractatus de computis et scripturis;" printed by Paganino de Paganini, Venice, 1494 (courtesy of Wellcome Collection).

[4] William A. Bernstein. Luca pacioli the father of accounting. In *The Air Force Comptroller*, volume 10(2) of *Air Force recurring publication 170-2*, pages 44–45. Office of the Comptroller, United States Air Force, 1976.

[5] Paolo Quattrone. Governing social orders, unfolding rationality, and Jesuit accounting practices. *Administrative Science Quarterly*, 60(3):411–445, 2015.

Pacioli's work represents a quantum leap in the realm of *descriptive inventory analytics*, a discipline that would evolve into a fundamental part of inventory management. However, no progress was made in the realm of *predictive* and *prescriptive inventory analytics* until late 1800, when Edgeworth,[6] in his "Mathematical Theory of Banking," used the central limit theorem to determine cash reserves needed to satisfy random withdrawals from depositors, thus embedding a *predictive* probabilistic model within a *prescriptive* mathematical model to support inventory control decisions.

[6] Francis Y. Edgeworth. The mathematical theory of banking. *Journal of the Royal Statistical Society*, 51(1):113–127, 1888.

From these early results, over the past 150 years, inventory control has evolved into an independent discipline. The aim of this book is to provide an introduction to this discipline.

After introducing the foundations of inventory systems, in chapter "Deterministic Inventory Control" we survey *prescriptive analytics* models for deterministic inventory control, in chapter "Demand Forecasting" we discuss *predictive analytics* techniques for demand forecasting in inventory control, which originate in the realm of time series analysis and forecasting. Finally, in chapters "Stochastic Inventory Control" and "Multi-echelon Inventory Systems" we survey *prescriptive analytics* models for stochastic inventory control.

Inventory Systems

https://doi.org/10.11647/OBP.0252.01

Introduction

In this chapter, we first discuss key reasons for keeping inventory in supply chain management, and strategies that can be adopted to review inventory. We then introduce a simple inventory system to motivate our discussion, we illustrate what costs need to be considered while controlling inventory, and the impact of a supplier lead time on the inventory system.

Topics

The role of inventory in supply chain management

Fig. 3 A warehouse.

A SUPPLY CHAIN is a system of organisations, people, technology, activities, information, and resources involved in moving a product or service from supplier to customer.

SUPPLY CHAIN MANAGEMENT is the management of this flow of products and services; it encompasses the movement and storage of raw materials, of work-in-process inventory, and of finished goods from point of origin to point of consumption. Inventory systems, such as warehouses (Fig. 3) and distribution centers (Fig. 4), are at the heart of supply chain management. In the rest of this chapter we focus on inventory systems and we discuss their nature.

THERE ARE THREE MAIN REASONS for keeping inventory: **time**, **uncertainty**, and **economies of scale**.

Reasons for keeping inventory

TIME. Moving goods along a supply chain is time consuming, e.g. after an order is placed, it usually takes time (**lead time**) to receive the goods. Inventory can be used to ensure business continuity during these delays. If the lead time is known and fixed, it can be addressed by ordering goods in advance. It is only when lead time is uncertain that inventory becomes essential.

UNCERTAINTY. Lead time, demand, supply, and other supply chain characteristics may be subject to uncertainty; inventory is then maintained as a buffer to hedge against this uncertainty.

ECONOMIES OF SCALE. A pure **lean** approach, i.e. "one unit at a time at a place where a user needs it, when (s)he needs it" typically incurs lots of costs in terms of logistics. Economies of scale can be pursued via bulk buying, movement, and storing; but they also come with inventory as a side effect.

INVENTORY counts as a current asset on the balance sheet because, in principle, it can be sold and turned into cash. However, inventory ties up money that could serve for other purposes. Moreover, it may require additional expense for its storage, e.g. warehouse rent, and protection, e.g. insurance. Inventory may also cause significant tax expenses, depending on particular countries' laws regarding depreciation of inventory.

INVENTORY REVIEW is the process by which a manager determines inventory quantities on hand. There are two main review strategies commonly adopted: **continuous review** and **periodic review**.

CONTINUOUS REVIEW operates by continuously recording receipt and disbursement for every item of inventory. This is an expensive and cumbersome strategy that is required for critical items (so-called A-type) to minimize costly machine shut-downs and customer complaints.

PERIODIC REVIEW requires a physical count (**stock take**) of goods on hand at the end of a period. This is a simple strategy that concentrates records and adjustments mostly at the end of a *period* (e.g. a week). It is widely used for items that are marginally important or less important than A-type ones (so-called B-type and C-type).[7]

[7] The process of classifying items into different categories on the basis of their importance is known as ABC analysis; for a survey on this topic see [van Kampen et al., 2012].

Fig. 4 A distribution center.

A simple inventory system

In what follows, we shall consider the simplest inventory system one may conceive (Fig. 5).

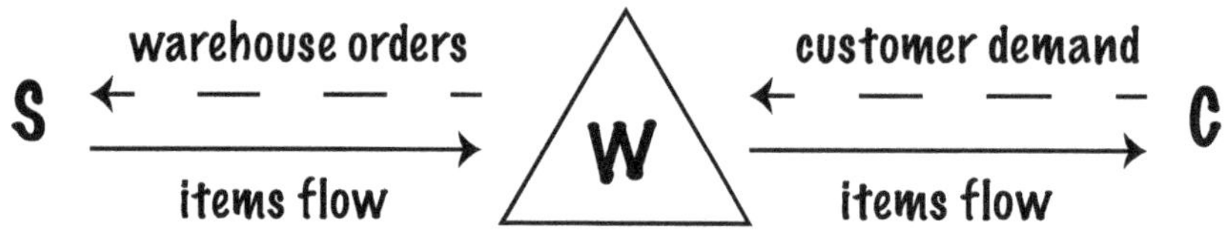

Fig. 5 A simple inventory system; physical flows and information flows are represented via solid and dashed lines, respectively.

THE SYSTEM comprises a warehouse (W) represented by means of a triangle, which in Total Quality Management (TQM) diagrams (Fig. 6) is a commonly adopted symbol to denote inventory/storage.

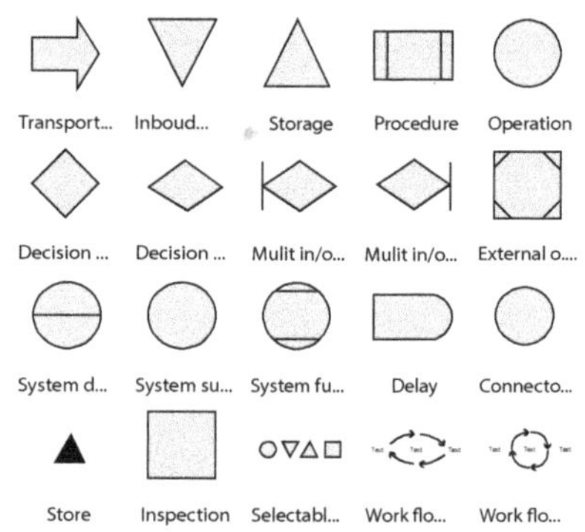

Fig. 6 TQM diagram shapes.

THE WAREHOUSE stocks a **single item type**, which in technical term we refer to as **stock keeping unit**, or SKU in short.

THE STATE of the warehouse is given by its **inventory level**.

THE WAREHOUSE faces **demand** from **customers** (C), and can satisfy this demand by issuing a sufficient number of items from its inventory.

THE WAREHOUSE can only meet demand if **on hand inventory** is large enough.

ITEMS can be **ordered** by the warehouse from a **supplier** (S) to maintain a suitable **inventory level**.

Simulating a simple inventory system in Python

We next discuss how to model a supplier, a warehouse and a customer in Python.

THE WAREHOUSE is shown in Listing 1. This class embeds a state variable `i` to **track the warehouse inventory level**. There are four methods: `order`, to **replenish inventory** by a quantity `Q`; `on_hand_inventory`, to **inspect the on hand inventory**;[8] `issue`, to **issue items from the warehouse** and meet `demand`; and `review_inventory`, to **review and keep an account of inventory** over time.

INVENTORY REVIEW is a key aspect of inventory management. In the code, method `review_inventory` is called before and after an order is issued in method `order`, and before and after inventory is issued to meet demand in method `issue`. The method features an argument `time`, to keep an account of the time at which the inventory level has been inspected. Note that when the method is called after inventory is issued to meet demand in method `issue`,

[8] Note that, while the inventory level `i` may go negative to account for backorders (i.e. orders that have not been satisfied yet due to lack of stock), the on hand inventory (the physical stock in the warehouse) is always nonnegative.

```
class Warehouse:
   def __init__(self, inventory_level):
      self.i = inventory_level
      self.review_inventory(0)

   def order(self, Q, time):
      self.review_inventory(time)
      self.i += Q
      self.review_inventory(time) # orders are received at the beginning of a period

   def on_hand_inventory(self):
      return max(0, self.i)

   def issue(self, demand, time):
      self.review_inventory(time)
      self.i = self.i-demand
      self.review_inventory(time+1) # demand is realised at the end of a period

   def review_inventory(self, time):
      try:
         self.levels.append([time, self.i])
         self.on_hand.append([time, self.on_hand_inventory()])
      except AttributeError:
         self.levels, self.on_hand = [[0, self.i]], [[0, self.on_hand_inventory()]]
```

Listing 1 A warehouse in Python.

the argument is `time+1`; this is because **we assume demand is observed at the end of a period**. In contrast, the argument is simply `time` when the method is called after an order is received in method `order`, because **we assume orders are received at the beginning of a period**, before any demand is observed.

THE CUSTOMER AND THE SUPPLIER do not need to be explicitly modelled as classes, since we will assume no lead time and an infinite supply available upon ordering, and a constant source of demand over our simulation horizon.

```
import matplotlib.pyplot as plt
import numpy as np
import pandas as pd

def plot_inventory(values, label):

   # data
   df=pd.DataFrame({'x':
         np.array(values)[:,0], 'fx':
         np.array(values)[:,1]})

   # plot
   plt.xticks(range(len(values)),
            range(1,len(values)+1))
   plt.xlabel("$t$")
   plt.ylabel("items")
   plt.plot( 'x', 'fx', data=df,
         linestyle='-', marker='o',
         label=label)
```

Listing 2 Plotting inventory in Python.

Example 1. *We simulate operations of this simple inventory system by leveraging the Python code in Listing 2 and Listing 3. The warehouse initial inventory is 100 units. The customer demand rate is 5 unit per period. We simulate $N = 20$ periods. The behaviour of the inventory level at the end of each period is shown in Fig. 7. The system starts with 100 units of inventory at the beginning of period 1; 5 units of inventory are consumed in every period; at the end of period 20 (or equivalently, at the beginning of period 21), the system inventory level is 0.*

```
initial_inventory = 100
w = Warehouse(initial_inventory)

demand_rate = 5 # customer demand rate per period

N = 20 # planning horizon length
for t in range(N):
   w.issue(demand_rate, t)

plot_inventory(w.levels, "inventory level")
plt.legend()
plt.show()
```

Listing 3 Simulating the behaviour of a warehouse in Python: the warehouse initial inventory level is 100, the customer demand rate is 5 units per period.

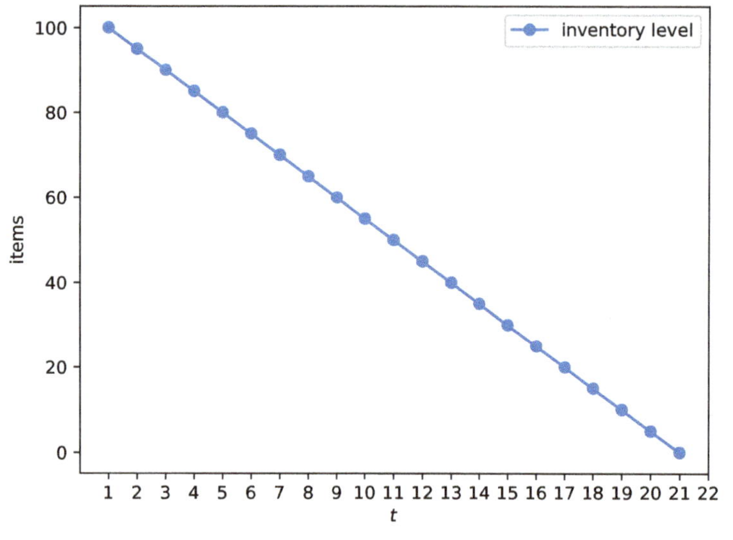

Fig. 7 Simulating the behaviour of a warehouse in Python: inventory level at the end of each period $t \in \{1, 20\}$ when the initial inventory level is 100.

THE SYSTEM we have just simulated is an example of **periodic review** inventory control. Inventory is reviewed at the end of each period,[9] after demand has been observed. Note that in this simulation, we have relied on the inventory available at the beginning of the planning horizon, and we have not issued any order.

[9] And hence also at the beginning of the next period, since these two instants coincides.

Alternatively, we can set `initial_inventory = 50` and issue an order of size 50 at the beginning of period 10. To do so, we slightly amend the code in Listing 3, by replacing the `for` loop as shown in Listing 4. Note that, in period 10, when we place an order, we review inventory (`levels.append([t, w.inventory_level()])`) both before and after ordering. Finally, in every period, as before we review inventory after demand has been observed, at the end of period t (or equivalently, at the beginning of period $t + 1$, as these two instants coincide).

Listing 4 Revised `for` loop.

```
for t in range(N):
  if(t == 10):
    w.order(50, t) # place an order of size 50 in period 10
  w.issue(demand_rate, t)
```

In Fig. 8 we plot the behaviour of the system under this new **control policy**, which meets demand in periods $1, \ldots, 10$ by leveraging the initial inventory, and meets demand in periods $11, \ldots, 20$, by means of an order of size 50 in period 10.

Let us now assume that the order in period 10 has size 40 (Fig. 9). The order is clearly not sufficient to cover demand until the end of the planning horizon. The **closing inventory level** at the end of period 19 and 20 is now **negative** and equal to -5 and -10, respectively: the system ran out of stock and we have observed a **stockout**.

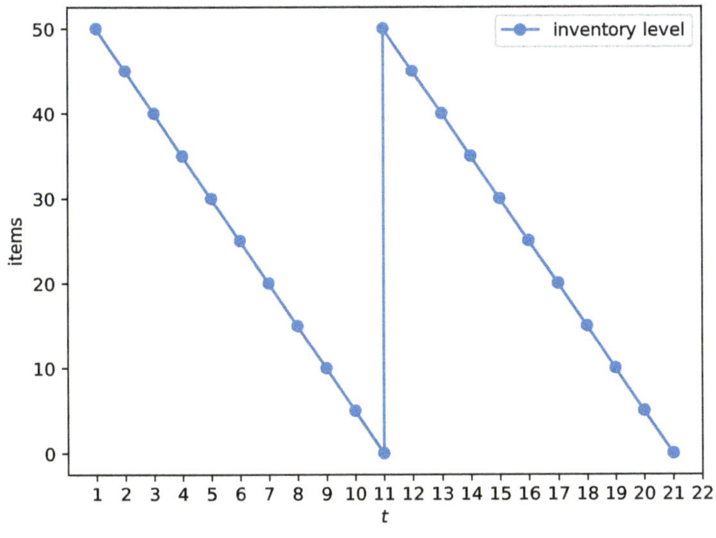

Fig. 8 Simulating the behaviour of a warehouse in Python: inventory level at the end of each period $t \in \{1, 20\}$ when the initial inventory level is 50 and an order of size 50 is placed in period 10.

When customer demand exceeds on hand inventory, the warehouse manager may decide to **lose** or **backorder** a sale. In the former case, we say that the inventory system operates under *lost sales*; in the latter case, we say that the inventory system operates under *backorders*. If a sale is backordered, the inventory level will go negative to keep track of pending demand, which will be met as soon as a suitable quantity is received from the supplier. An inventory system may backorder all (full/complete backorders) or only a part (partial backorders) of the demand that exceeds on hand inventory.

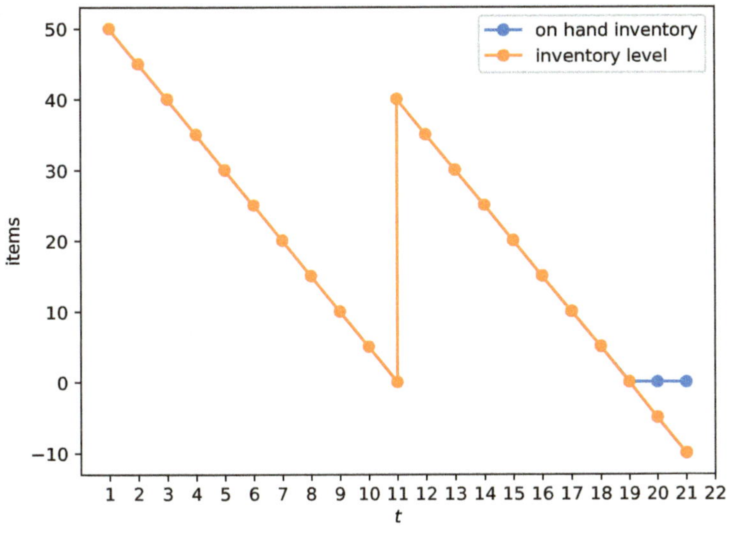

Fig. 9 Simulating the behaviour of a warehouse in Python: inventory level at the end of each period $t \in \{1, 20\}$ when the initial inventory level is 50 and an order of size 40 is placed in period 10. Observe that, while the on hand inventory is zero at the end of period 19 and 20, the inventory level is negative.

Inventory costs

Consider a warehouse facing customer demand at a constant rate of five units per period over a potentially infinite time horizon. Should we keep inventory? If so, how much? It is not possible to provide an answer to these two questions without further information on the costs the warehouse manager faces, and on other operating characteristics of the inventory system.

For instance, if every time the warehouse manager issues an order of size $Q > 0$ to its supplier, the only cost involved in the transaction is the **per unit purchase cost** v of an item; and if the supplier delivers the quantity Q immediately — i.e. **no delivery lead time** — then it is clear the warehouse manager should adopt a **lean control strategy**: no inventory should be kept, and whenever a demand unit materialises from the customer, the warehouse manager should simply order one unit from the supplier and meet the customer demand from the order quantity immediately received. Assuming a **selling price** $p > v$, the system would generate a profit $p - c$ for every unit of demand met. The behaviour of a lean warehouse in Python can be simulated via the code in Listing 5 and it is shown in Fig. 10.

```
initial_inventory = 0
w = Warehouse(initial_inventory)

demand_rate = 5

N = 20 # planning horizon length
for t in range(N):
   w.order(5, t) # place an order of
        size 5
   w.issue(demand_rate, t)

plot_inventory(w.levels, "inventory
     level")
plt.legend()
plt.show()
```

Listing 5 Simulating the behaviour of a lean warehouse in Python: the warehouse initial inventory level is 0, the customer demand rate is 5 units per period, and orders of size 5 are issued in every period. The total demand over the 20-period planning horizon is 100 units, the system would therefore generate a profit of $100(p - c)$.

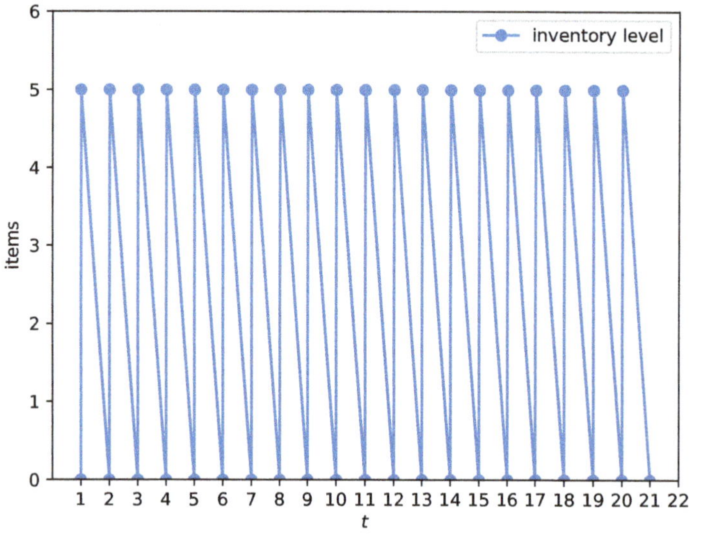

Fig. 10 Simulating the behaviour of a lean warehouse in Python.

In essence, in a lean inventory system, the inventory manager does away with inventory and, without holding any stock, acts as an intermediary. The TQM diagram in Fig. 5 therefore takes the new form shown in Fig. 11: the physical inventory originally represented as a triangle has been replaced by an ordering and demand fulfilment process, represented via a rectangle with rounded edges.

Lean inventory management.

Fig. 11 A lean inventory system.

THE PER UNIT PURCHASE COST is hardly the only cost involved in managing a warehouse. We next summarise other cost factors that are often encountered by inventory managers.

Inventory cost factors

INVENTORY REVIEW COST. This cost is charged when a physical inventory inspection takes place.

FIXED ORDERING COST. Independent of the size of the order, it is charged every time an order is issued.

PER UNIT PURCHASE COST. This is a cost that is proportional to the number of items that are ordered.

PER UNIT INVENTORY HOLDING COST. This is a cost that is charged for every unit carried forward in stock from one period to the next in the planning horizon.

FIXED STOCKOUT/BACKORDER PENALTY COST. This is a cost that is charged every time the inventory level turns negative; the associated units of demand that drove the inventory level negative may be lost or backordered; this cost is independent of the magnitude of the stockout observed.

PER UNIT STOCKOUT/BACKORDER PENALTY COST. This is a cost charged when a unit of demand is backordered or lost; it is charged once, and if the unit of demand is backordered, it is independent of the time it takes to fulfil it.

PER UNIT, PER TIME PERIOD STOCKOUT/BACKORDER PENALTY COST. This is a cost that is charged when a unit of demand is backordered; it is charged for every time period the unit of demand remains short.

Lot sizing

If there are **fixed costs** — independent of the order quantity Q — for placing an order with the supplier, e.g. cost of dispatching a truck, and if we incur **inventory holding costs**, then we enter the realm of **lot sizing**, and it becomes necessary to hold inventory.

Deterministic supplier lead time

Consider the case in which the supplier is not able to deliver the order quantity Q immediately, but will be able to deliver it after a known and fixed lead time. What should we do?

Fig. 12 An inventory system subject to supplier lead time.

If it takes a 1-period **lead time** to receive an order from the supplier, and we assume that all demand must be immediately satisfied from on hand inventory, this means that at the beginning of the planning horizon we must already hold at least 5 items in stock — or equivalently we should expect to receive five correctly timed outstanding orders from the supplier — otherwise the problem would not admit a solution.

Moreover, as soon as we observe the first unit of demand, we will have to immediately issue an order to replace the item we have just sold. As in lean inventory management, under this strategy, **we do not need to hold inventory in the warehouse**; this is because the inventory we need to run the system takes the form of **inbound outstanding orders**, that is orders yet to be received from our supplier — in TQM diagram notation, these are represented via an inverted triangle (Fig. 12). The on hand inventory plus any outstanding order, minus backorders, is a quantity called the **inventory position**; as we will see, this quantity, which keeps track of outstanding orders, will be useful to control our system.

To model an inventory system subject to deterministic supplier lead time we must adopt a different strategy from that which we previously followed. More specifically, we will adopt a **Discrete Event Simulation** (DES) approach.

```
from queue import PriorityQueue

events = PriorityQueue()

events.put((0.3, "Customer demand"))
events.put((0.5, "Customer demand"))
events.put((0, "Order"))
events.put((0.10, "Customer demand"))

while events:
    print(events.get())

# Will print events in the order:
# (0, 'Order')
# (0.1, 'Customer demand')
# (0.3, 'Customer demand')
# (0.5, 'Customer demand')
```

Listing 6 A priority queue in Python.

A PRIORITY QUEUE is an abstract data structure (a data structure defined by its behaviour) that is like a normal queue, but where each item has a special **key** to quantify its **priority**. For instance, airlines may give luggage on the conveyer belt based on the status or ticket class of the passengers. Baggage tagged with *priority* or *business* or *first-class* usually arrives earlier than other non-tagged baggage. In Listing 6 we implement a simple priority queue.

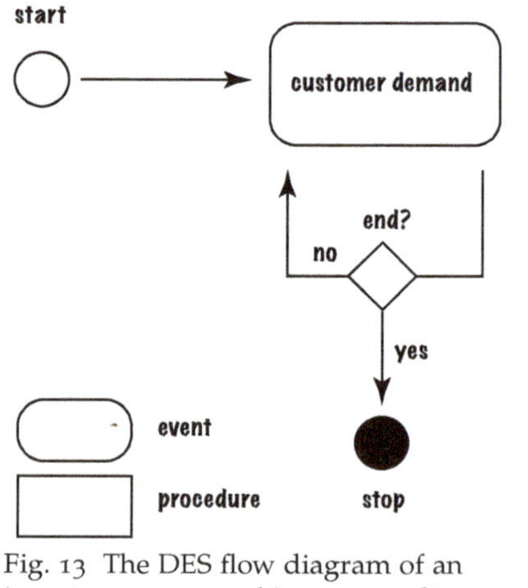

Fig. 13 The DES flow diagram of an inventory system subject to supplier lead time.

TO MODEL OUR DES SYSTEM, in Listing 7 we extend the behaviour of `Warehouse` to model both inventory level and inventory position. A general-purpose DES loop implementing the flow diagram[10] in Fig. 13 is shown in Listing 8. The method `start` enters a `while` loop that repeatedly extracts events from a priority queue and executes them until the end of the simulation horizon. The method `schedule` schedules an event after `time_lag`.

[10] Arnold H. Buss. A tutorial on discrete-event modeling with simulation graphs. In C. Alexopoulos, I Kang, W. R. Lilegdon, and D. Goldsman, editors, *Proceedings of the 1995 Winter Simulation Conference ed.*, Arlington, Virginia, 1995.

Listing 7 The extended `Warehouse` class that models both inventory level and inventory position.

```
class Warehouse:
   def __init__(self, inventory_level):
      self.i = inventory_level
      self.o = 0 # outstanding_orders

   def receive_order(self, Q, time):
      self.review_inventory(time)
      self.i, self.o = self.i + Q, self.o - Q
      self.review_inventory(time)

   def order(self, Q, time):
      self.review_inventory(time)
      self.o += Q
      self.review_inventory(time) # orders are received at the beginning of a period

   def on_hand_inventory(self):
      return max(0,self.i)

   def issue(self, demand, time):
      self.review_inventory(time)
      self.i = self.i-demand
      self.review_inventory(time+1) # demand is realised at the end of a period

   def inventory_position(self):
      return self.o+self.i

   def review_inventory(self, time):
      try:
         self.levels.append([time, self.i])
         self.on_hand.append([time, self.on_hand_inventory()])
         self.positions.append([time, self.inventory_position()])
      except AttributeError:
         self.levels, self.on_hand = [[0, self.i]], [[0, self.on_hand_inventory()]]
         self.positions = [[0, self.inventory_position()]]
```

To model an inventory system subject to deterministic supplier lead time, we will create different classes of events: orders, demand, etc. We start with a generic `EventWrapper` in Listing 8, which is then specialised into a `CustomerDemand` event in Listing 9.

Listing 8 A DES engine in Python.

```
from queue import PriorityQueue

class EventWrapper():
   def __init__(self, event):
      self.event = event

   def __lt__(self, other):
      return self.event.priority < other.event.priority

class DES():
   def __init__(self, end):
      self.events, self.end, self.time = PriorityQueue() , end, 0

   def start(self):
      while True:
         event = self.events.get()
         self.time = event[0]
         if self.time < self.end:
            event[1].event.end()
         else:
            break

   def schedule(self, event: EventWrapper, time_lag: int):
      self.events.put((self.time + time_lag, event))
```

The `CustomerDemand` event in Listing 9, when invoked via the method `end`, generates a customer demand of 5 units at the warehouse. Finally, the event reschedules itself with a delay of 1 time period.

Listing 9 The `CustomerDemand` event.

```
class CustomerDemand:
    def __init__(self, des: DES, demand_rate: float, warehouse: Warehouse):
        self.d = demand_rate # the demand rate per period
        self.w = warehouse # the warehouse
        self.des = des # the Discrete Event Simulation engine
        self.priority = 1 # denotes a low priority

    def end(self):
        self.w.issue(self.d, self.des.time)
        self.des.schedule(EventWrapper(self), 1)
```

Listing 10 Simulating the behaviour of a warehouse via DES in Python: the warehouse initial inventory level is 100, the customer demand rate is 5 units per period.

```
initial_inventory = 100
w = Warehouse(initial_inventory)

N = 20 # planning horizon length
des = DES(N)

d = CustomerDemand(des, 5, w)
des.schedule(EventWrapper(d), 0) # schedule a demand immediately
des.start()

plot_inventory(w.positions, initial_inventory + 10, "inventory position")
plot_inventory(w.levels, initial_inventory + 10, "inventory level")
plt.legend()
plt.show()
```

In Listing 10 we show how to simulate the behaviour of a warehouse via DES in Python for our previous numerical example. The warehouse initial inventory level is 100, the customer demand rate is 5 units per period, and the system is simulated for $N = 20$ periods. The result of the simulation is shown in Fig. 14 and, as expected, it is identical to Fig. 7.

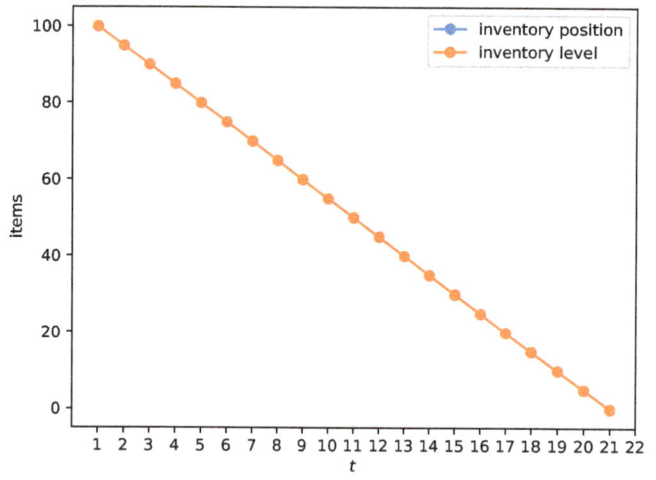

Fig. 14 Simulating the behaviour of a warehouse in Python: inventory level at the end of each period $t \in \{1, 20\}$ when the initial inventory level is 100.

Let us now reduce the initial inventory level to 50, and schedule an order at time 10, for which the delivery lead time is 1 period. To model an order, we create an `Order` event, to capture the delivery lead time, we create a `ReceiveOrder` event (Listing 11) that is triggered by the `Order` event with a delay of `lead_time` periods.

```
class Order:
    def __init__(self, des: DES, Q: float, warehouse: Warehouse, lead_time: float):
        self.Q = Q # the order quantity
        self.w = warehouse # the warehouse
        self.des = des # the Discrete Event Simulation engine
        self.lead_time = lead_time
        self.priority = 0 # denotes a high priority

    def end(self):
        self.w.order(self.Q, self.des.time)
        self.des.schedule(EventWrapper(ReceiveOrder(self.des, self.Q, self.w)),
            self.lead_time)

class ReceiveOrder:
    def __init__(self, des: DES, Q: float, warehouse: Warehouse):
        self.Q = Q # the order quantity
        self.w = warehouse # the warehouse
        self.des = des # the Discrete Event Simulation engine
        self.priority = 0 # denotes a high priority

    def end(self):
        self.w.receive_order(self.Q, self.des.time)
```

Listing 11 The `Order` event and the `ReceiveOrder` event.

Finally, we set up the system and we simulate it via the code in Listing 12. Fig. 15 illustrates the behaviour of the inventory level and of the inventory position for this system. Because of the order lead time, to prevent stockouts, it is necessary to anticipate the ordering time by 1 period, which is the very same length of the lead time.

```
N, initial_inventory = 20, 50 # planning horizon length and initial inventory
w, des = Warehouse(initial_inventory), DES(N)

d = CustomerDemand(des, 5, w)
des.schedule(EventWrapper(d), 0) # schedule a demand immediately
o = Order(des, 50, w, 1)
des.schedule(EventWrapper(o), 9) # schedule an order of size 50 in period 9
des.start()

plot_inventory(w.positions, "inventory position")
plot_inventory(w.levels, "inventory level")
plt.legend()
plt.show()
```

Listing 12 Simulating the behaviour of a warehouse via DES in Python: the warehouse initial inventory level is 50, the customer demand rate is 5 units per period, an order is scheduled at time 9, and order the lead time is 1.

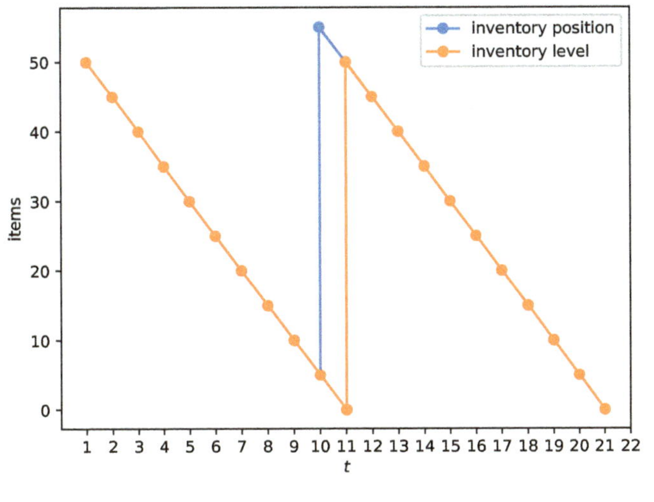

Fig. 15 Simulating the behaviour of a warehouse in Python: inventory level and inventory position at the end of each period $t \in \{1, 20\}$ when the initial inventory level is 50, an order is scheduled at time 9, and order the lead time is 1.

Deterministic Inventory Control

https://doi.org/10.11647/OBP.0252.02

Introduction

In this chapter, we discuss inventory control in a deterministic setting. We first discuss the cost factors that should be considered, and we show how to model and simulate the system running costs. We finally introduce prescriptive analytics models to determine the economic lot size under a variety of settings.

Topics

Accounting for costs

The simplest lot sizing instance one may conceive includes two cost factors: **a fixed ordering cost** (K), which is charged every time an order is issued, and it is a cost that is independent of the size of the order; and a **per unit inventory holding cost** (h), which is charged for every unit carried forward in stock from one period to the next in the planning horizon. In the first instance, we will assume that all demand must be met, hence the per unit item purchase cost can be ignored for all practical purposes. The revised `Warehouse` class is shown in Listing 13.

Listing 13 The extended `Warehouse` class that models costs.

```
from collections import defaultdict

class Warehouse:
   def __init__(self, inventory_level, fixed_ordering_cost, holding_cost):
      self.i, self.K, self.h = inventory_level, fixed_ordering_cost, holding_cost
      self.o = 0 # outstanding_orders
      self.period_costs = defaultdict(int) # a dictionary recording cost in each
           period

   def receive_order(self, Q, time):
      self.review_inventory(time)
      self.i, self.o = self.i + Q, self.o - Q
      self.review_inventory(time)

   def order(self, Q, time):
      self.review_inventory(time)
      self.period_costs[time] += self.K # incur ordering cost and store it in a
           dictionary
      self.o += Q
      self.review_inventory(time)

   def on_hand_inventory(self):
      return max(0,self.i)

   def issue(self, demand, time):
      self.review_inventory(time)
      self.i = self.i-demand

   def inventory_position(self):
      return self.o+self.i

   def review_inventory(self, time):
      try:
         self.levels.append([time, self.i])
         self.on_hand.append([time, self.on_hand_inventory()])
         self.positions.append([time, self.inventory_position()])
      except AttributeError:
         self.levels, self.on_hand = [[0, self.i]], [[0, self.on_hand_inventory()]]
         self.positions = [[0, self.inventory_position()]]

   def incur_holding_cost(self, time): # incur holding cost and store it in a
        dictionary
      self.period_costs[time] += self.on_hand_inventory()*self.h
```

To account for costs incurred by carrying over inventory from one period to the next we need to define an `EndOfPeriod` event (Listing 14) that is scheduled for the first time at the end of the first period, and which reschedules itself to occur at the end of every subsequent period.

```
class EndOfPeriod:
    def __init__(self, des: DES, warehouse: Warehouse):
        self.w = warehouse # the warehouse
        self.des = des # the Discrete Event Simulation engine
        self.priority = 2 # denotes a low priority

    def end(self):
        self.w.incur_holding_cost(self.des.time)
        self.des.schedule(EventWrapper(EndOfPeriod(self.des, self.w)), 1)
```

Listing 14 The `EndOfPeriod` event to record inventory holding costs.

Example 2. *We simulate operations of a simple inventory system by leveraging the Python code in Listing 15. The warehouse initial inventory is 0 units. The customer demand rate is 10 unit per period. We simulate $N = 20$ periods. We order 50 units in periods 1, 5, 10, and 15; the delivery lead time is 0 periods (i.e. no lead time). The fixed ordering cost is 100, the per unit inventory holding cost is 1. After simulating the system, we find that the average cost per unit time is 40; costs incurred in each period are shown in Table 1.*

```
instance = {"inventory_level": 0, "fixed_ordering_cost": 100, "holding_cost": 1}
w = Warehouse(**instance)

N = 20 # planning horizon length
des = DES(N)

d = CustomerDemand(des, 10, w)
des.schedule(EventWrapper(d), 0) # schedule a demand immediately

lead_time = 0
o = Order(des, 50, w, lead_time)
for t in range(0,20,5):
    des.schedule(EventWrapper(o), t) # schedule orders
des.schedule(EventWrapper(EndOfPeriod(des, w)), 0) # schedule EndOfPeriod
    immediately

des.start()

print("Period costs: "+str([w.period_costs[e] for e in w.period_costs]))
print("Average cost per period: "+ '%.2f' % (sum([w.period_costs[e] for e in
    w.period_costs])/len(w.period_costs)))

plot_inventory(w.positions, "inventory position")
plot_inventory(w.levels, "inventory level")
plt.legend()
plt.show()
```

Listing 15 Simulating the behaviour of a warehouse in Python: inventory level and inventory position at the end of each period $t \in \{1, 20\}$ when the initial inventory level is 0; orders are scheduled periods 1, 5, 10, and 15; and order the lead time is 0. The fixed ordering cost is 100, the per unit inventory holding cost is 1.

Period	1	2	3	4	5	6	7	8	9	10
Cost	140	30	20	10	0	140	30	20	10	0

Period	11	12	13	14	15	16	17	18	19	20
Cost	140	30	20	10	0	140	30	20	10	0

Table 1 Costs incurred in each period $t \in \{1, 20\}$ when the initial inventory level is 0, orders are scheduled every 5 periods, and order the lead time is 0. The fixed ordering cost is 100, the per unit inventory holding cost is 1.

The Economic Order Quantity

Consider an inventory system subject to a constant rate of d units of demand per time period. We shall assume that inventory is continuously reviewed (continuous review) and that the ordering/production process is instantaneous, i.e. as soon as we order a product or a batch of products, we immediately receive it. The order quantity can take any nonnegative real value. There is a proportional holding cost h per unit per time period for carrying items in stock. All demand must be met on time, i.e. it cannot be backordered.

In absence of fixed costs associated with issuing an order or with setting up production, since we face inventory holding costs, it is clear that the best strategy to meet demand is to order/produce a product as soon as demand for it materialises: a pure reactive and lean strategy. In practice, however, firms do face fixed production/setup costs. In this case, the optimal control strategy is less obvious.

The problem of determining the "economic" order quantity[11] (EOQ) in presence of fixed and variable production costs as well as proportional inventory holding cost was first studied by Harris at the beginning of the last century.[12] For a historical perspective see [Erlenkotter, 1990]. Harris' original "manufacturing quantity curves" are shown in Fig. 16.

The elements of the problem are summarized in Listing 16.

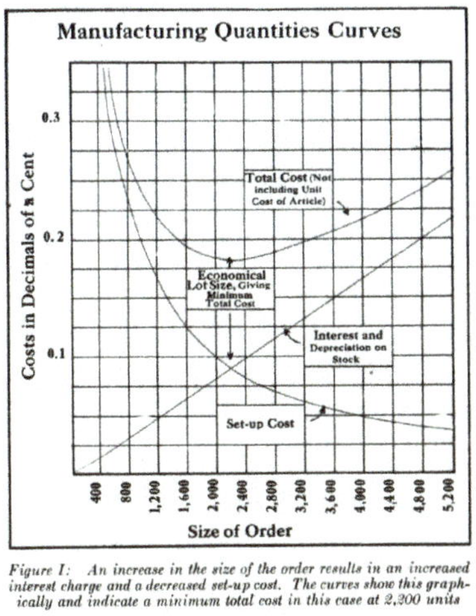

Fig. 16 Harris' manufacturing quantity curves from [Harris, 1913] (Courtesy of HathiTrust).

[11] Economic is used as a synonym of optimal.

[12] Ford W. Harris. How many parts to make at once. *Factory, The Magazine of Management*, 10(2):135–136, 1913.

Listing 16 The eoq class.

```
class eoq:
  def __init__(self, K: float, h: float, d: float, v: float):
    """
    Constructs an instance of the Economic Order Quantity problem.

    Arguments:
      K {float} -- the fixed ordering cost
      h {float} -- the proportional holding cost
      d {float} -- the demand per period
      v {float} -- the unit purchasing cost
    """
    self.K, self.h, self.d, self.v = K, h, d, v
```

In the EOQ, the demand is constant, we operate under continuous review, and backorders are not allowed; hence, the following property ensures one does not incur unnecessary holding costs.

Lemma 1 (Zero inventory ordering). *Given an order quantity Q it is optimal to issue an order as soon as the inventory level is zero.*

The inventory level as a function of time is shown in Fig. 17: as soon as inventory level hits zero, an order of size Q is immediately received and inventory immediately starts decreasing at rate d unit per period; the cycle repeats when inventory level hits zero again.

Definition 1 (Replenishment cycle). *A replenishment cycle is the time interval comprised within two consecutive orders.*

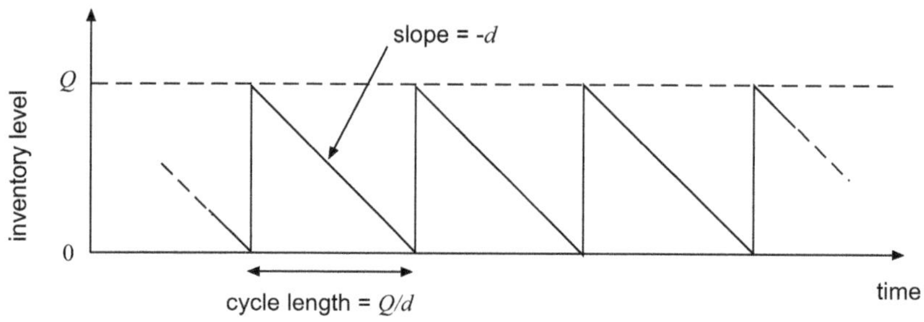

Fig. 17 EOQ "sawtooth" inventory curve.

Lemma 2 (Cycle length). *The length of an EOQ replenishment cycle is*

$$Q/d.$$

This is also known as the demand "coverage."

Consider a replenishment cycle of length R periods, a demand rate of d units/period and an order quantity $Q = dR$, which covers exactly the demand over R periods.

Lemma 3 (Average inventory level). *The average inventory level over the cycle is $Q/2$.*

Proof.

$$\frac{\int_0^R (Q - dr)\mathrm{d}r}{R}$$
$$= \frac{d}{Q}\left[rQ - \frac{dr^2}{2}\right]_0^{Q/d}$$
$$= \frac{d}{Q}\left[\left(\frac{Q^2}{d} - \frac{dQ^2}{2d^2}\right) - (0Q - 0)\right] = Q/2.$$

□

A key metric generally used to gauge inventory system performance is the so-called Implied Turnover Ratio.

Definition 2 (Implied Turnover Ratio). *The Implied Turnover Ratio (ITR) represents the number of times inventory is sold or used in a time period; this is expressed as average demand over average inventory*

$$2d/Q.$$

This information is important because it measures how fast a company is selling inventory and can be compared against industry benchmarks.

Cost analysis

The total cost of a strategy that issues an order of size Q as soon as the inventory level reaches zero can be expressed in terms of ordering and holding cost per replenishment cycle

$$C(Q) = \underbrace{\frac{K}{Q/d} + dv}_{\text{ordering cost}} + \underbrace{h\frac{Q}{2}}_{\text{holding cost}}. \qquad (1)$$

Since we operate under an infinite horizon and we assume that all demand must be met, in our cost analysis we can safely ignore the variable purchasing cost dv, which is constant and independent of Q, and consider the total "relevant cost" $C_r(Q) = C(Q) - dv$. These concepts are implemented in Listing 17.

Listing 17 EOQ cost functions in Python.

```
class eoq:
    def cost(self, Q: float) -> float:
        return self.fixed_ordering_cost(Q) + self.variable_ordering_cost(Q) +
            self.holding_cost(Q)

    def relevant_cost(self, Q: float) -> float:
        return self.fixed_ordering_cost(Q) + self.holding_cost(Q)

    def fixed_ordering_cost(self, Q: float) -> float:
        K, d = self.K, self.d
        return K/(Q/d)

    def variable_ordering_cost(self, Q: float) -> float:
        d, v = self.d, self.v
        return d*v

    def holding_cost(self, Q: float) -> float:
        h = self.h
        return h*Q/2
```

In Fig. 18 we plot the different components that make up the EOQ cost function as well as $C_r(Q)$.

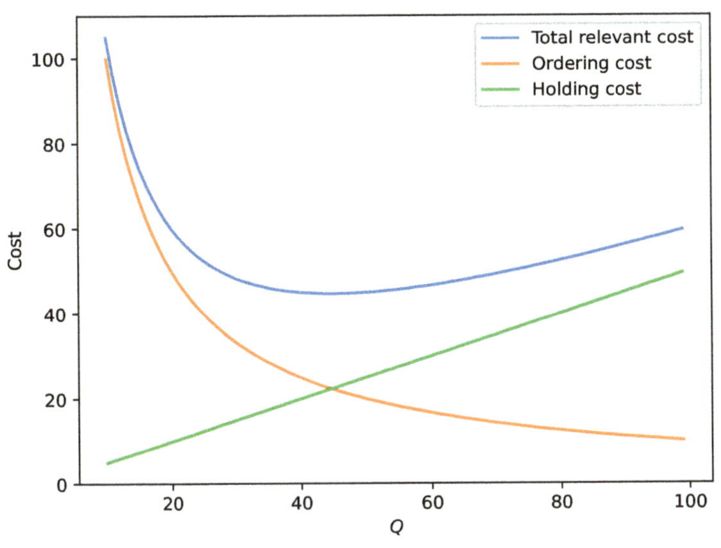

Fig. 18 EOQ cost functions.

Lemma 4 (Convexity of relevant cost). *$C_r(Q)$ is convex.*

Proof.

$$\frac{d^2 C_r(Q)}{dQ} = \frac{2Kd}{Q^3} > 0.$$

□

Optimal solution

Since $C(Q)$ is convex, its global minimum can be found via global optimisation approaches readily available in software libraries such as Python `scipy`. For instance, one may exploit Nelder-Mead[13] algorithm as shown in Listing 18.

[13] John A. Nelder and Roger Mead. A simplex method for function minimization. *The Computer Journal*, 7(4):308–313, 1965.

Listing 18 Compute Q^*.

```
from scipy.optimize import minimize

class eoq:
  def compute_eoq(self) -> float:
        x0 = 1 # start from a positive EOQ
        res = minimize(self.relevant_cost, x0, method='nelder-mead',
              options={'xtol': 1e-8, 'disp': False})
        return res.x[0]
```

The analytical closed-form optimal solution to the EOQ problem, the so-called Economic Order Quantity Q^* is shown in the following Lemma.

Lemma 5 (Economic Order Quantity).

$$Q^* = \sqrt{2Kd/h}. \tag{2}$$

Proof. By exploiting convexity of $C_r(Q)$, one sets its first derivative to zero

$$-\frac{Kd}{Q^2} + \frac{h}{2} = 0$$

and obtains a closed form for the optimal order quantity. □

The particular form of Q^* allows us to make some observations: as K increases we will issue larger orders; as h increases holding inventory becomes more expensive and we order more frequently; finally, as d increases the order quantity increases.

Lemma 6 (Relevant cost of ordering the Economic Order Quantity).

$$C_r(Q^*) = \sqrt{2Kdh} \tag{3}$$

Proof. This is obtained by plugging Eq. 2 into $C_r(Q)$. □

Example 3. *We consider the numerical example in Listing 19. Note that this is the same instance considered in Example 2. After running the code we obtain $Q^* = 44.72$ and $C_r(Q^*) = 44.72$. The replenishment cycle length is therefore $Q^*/d = 4.472$ periods.*

Listing 19 Numerical example 3.

```
instance = {"K": 100, "h": 1, "d": 10, "v": 2}
pb = eoq(**instance)
Qopt = pb.compute_eoq()
print("Economic order quantity: " + '%.2f' % Qopt)
print("Total relevant cost: " + '%.2f' % pb.relevant_cost(Qopt))
```

The fact that $Q^* = C_r(Q^*)$ is a direct consequence of Lemma 6 and $h = 1$.

Since we are operating under continuous review, although the instance parameters are the same, the cost obtained by applying the EOQ formula to the previous example is not directly comparable to that obtained via the simulation presented in Listing 15, which operates under periodic review. To address this issue, we need to adopt a finer discretisation of the planning horizon. This is shown in Listing 20 and in Listing 21. The cost of the simulated solution is now 44.78, which is equivalent to that obtained from the analytical solution. The behaviour of inventory over time is shown in Fig. 19.

```
instance = {"inventory_level": 0, "fixed_ordering_cost": 100,
   "holding_cost": 1.0/100} # holding cost rescaled
w = Warehouse(**instance)

des = DES()

demand = 10/100 # discretise each period into 100 periods
Q = 44.72 # optimal EOQ order quantity
N = round(Q/demand)*10 # planning horizon length: simulate 10 replenishment cycles
des.schedule(EventWrapper(EndOfSimulation(des, w)), N) # schedule EndOfSimulation

d = CustomerDemand(des, demand, w)
des.schedule(EventWrapper(d), 0) # schedule a demand immediately

lead_time = 0
o = Order(des, Q, w, lead_time)
for t in range(0, N, round(Q/demand)):
   des.schedule(EventWrapper(o), t) # schedule an order at the beginning of each
        replenishment cycle
des.schedule(EventWrapper(EndOfPeriod(des, w)), 1) # schedule EndOfPeriod at the
     end of period 1

des.start()

print("Period costs: "+str([w.period_costs[e] for e in w.period_costs]))
print("Average cost per period: "+ '%.2f' % (100*sum([w.period_costs[e] for e in
     w.period_costs])/len(w.period_costs)))

plot_inventory_finer(w.positions, "inventory position")
plot_inventory_finer(w.levels, "inventory level")
plt.legend(loc=1)
plt.show()
```

Listing 20 Simulating the behaviour of a warehouse in Python: DES simulated EOQ solution under a finer discretisation of the simulation horizon (100 smaller period for each original period).

```
def plot_inventory(values, label):

  # data
  df=pd.DataFrame({'x':
        np.array(values)[:,0], 'fx':
        np.array(values)[:,1]})

  # plot
  plt.xticks(range(0,len(values),200),
        range(0,len(values)//100,2))
        # a tick every 200 periods
  plt.xlabel("t")
  plt.ylabel("items")
  plt.plot( 'x', 'fx', data=df,
        linestyle='-', marker='',
        label=label)
```

Listing 21 Method plot_inventory under a finer discretisation of the simulation horizon (100 smaller period for each original period).

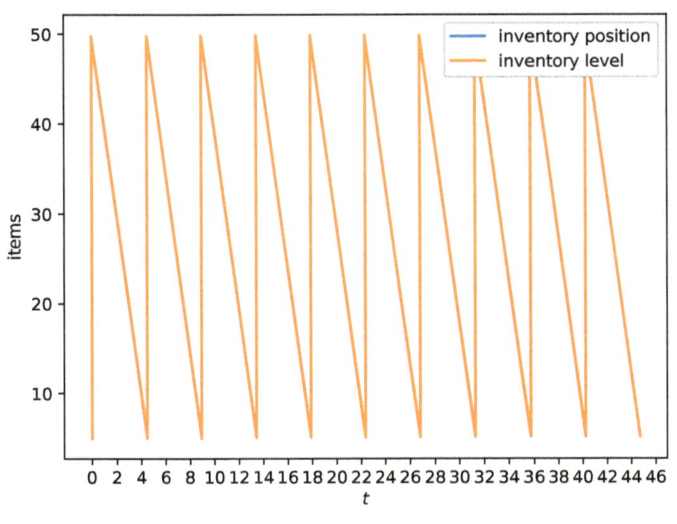

Fig. 19 Simulating the behaviour of a warehouse in Python under a finer discretisation of the simulation horizon (100 smaller period for each original period).

Sensitivity to variations of Q

Suppose management decides to order a quantity Q that differs from Q^*. The following Lemma is important in order to understand the sensitivity of the relevant cost to the choice of Q.

Lemma 7 (Sensitivity to Q^*). *Let* $Q > 0$

$$\frac{C_r(Q)}{C_r(Q^*)} = \frac{1}{2}\left(\frac{Q^*}{Q} + \frac{Q}{Q^*}\right) \tag{4}$$

Proof.

$$\begin{aligned}\frac{C_r(Q)}{C_r(Q^*)} &= \frac{Kd}{Q\sqrt{2Kd/h}} + h\frac{Q}{2\sqrt{2Kd/h}}\\ &= \frac{1}{2Q}\sqrt{\frac{2Kd}{h}} + h\frac{Q}{2}\sqrt{\frac{h}{2Kd}}\\ &= \frac{1}{2}\left(\frac{Q^*}{Q} + \frac{Q}{Q^*}\right).\end{aligned}$$

□

```
class eoq:
    def sensitivity_to_Q(self, Q:
         float) -> float:
        Qopt = self.compute_eoq()
        return 0.5*(Qopt/Q+Q/Qopt)
```

Listing 22 Compute sensitivity to variations of Q from Q^*.

Sensitivity can be computed as shown in Listing 22.

There are two key observations: the sensitivity of the relevant cost to the choice of Q only depends on Q and Q^*, not on the specific values of problem parameters K and h; moreover, this sensitivity is low.

Example 4. *In Fig. 20 we plot Eq. 4 for the numerical example presented in Listing 19. We can see that a difference of 10 units between* $Q^* = 44.72$ *and* $Q = 34.72$ *only leads to a 3.22% cost increase.*

Fig. 20 EOQ sensitivity to variations of Q from Q^*.

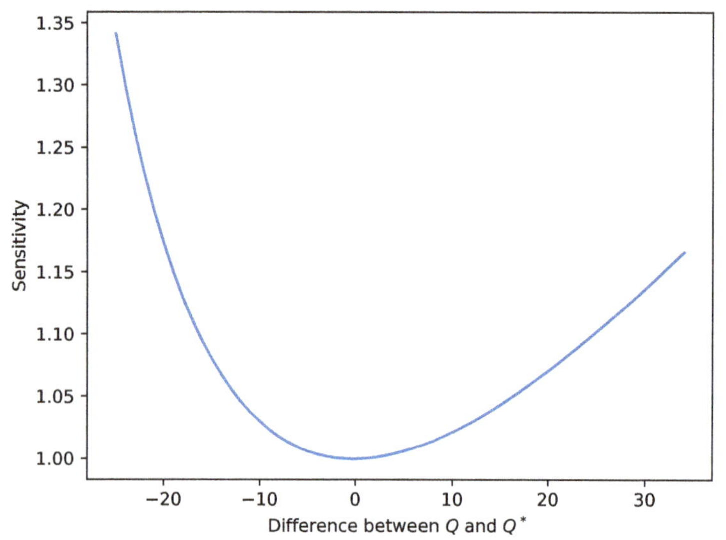

Incorrect estimation of fixed ordering and holding costs

Suppose we incorrectly estimate the fixed ordering cost K as K', by leveraging once more Eq. 4 we obtain

$$\frac{C_r(Q')}{C_r(Q^*)} = \frac{1}{2}e\left(\sqrt{\frac{K'}{K}}\right)$$

which implies that the cost of overestimating K is lower than that of underestimating it. A similar analysis can be carried out for the holding cost parameter; for which, however, the situation is reversed: $C_r(Q')/C_r(Q^*) = 0.5\sqrt{h/h'}$.

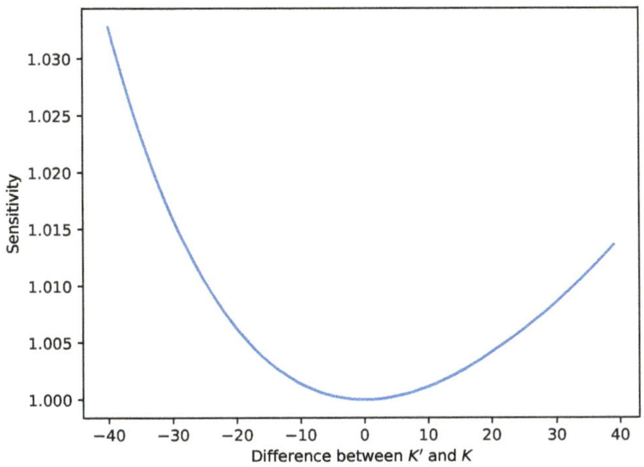

Fig. 21 EOQ sensitivity to K.

```
class eoq:
 def sensitivity_to_K(self, K: float)
        -> float:
        e = lambda x : x + 1/x
        return 0.5*(e(np.sqrt(K/self.K)))

    def sensitivity_to_h(self, h:
         float) -> float:
        e = lambda x : x + 1/x
        return 0.5*(e(np.sqrt(self.h/h)))
```

Listing 23 Compute sensitivity to estimation errors for K and h.

Sensitivity to K and h can be computed as shown in Listing 23.

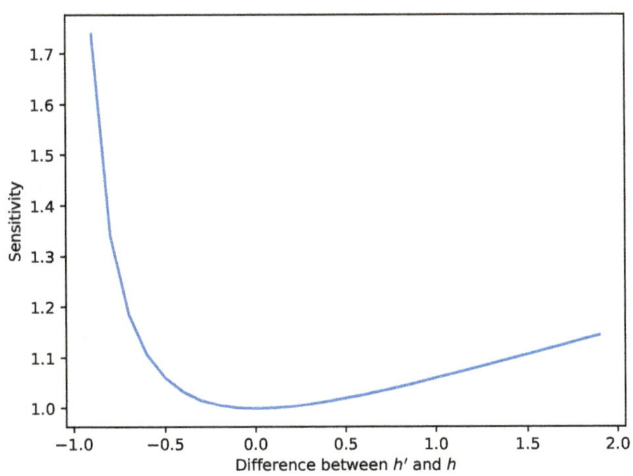

Fig. 22 EOQ sensitivity to h.

Example 5. *In the numerical example presented in Listing 19, if we underestimate K by 40% we underestimate $C_r(Q^*)$ by $0.5e(\sqrt{0.6}) = 0.0328$, i.e. 3.28%; if we overestimate K by 40% we overestimate $C_r(Q^*)$ by $0.5e(\sqrt{1.4}) = 0.0141$, i.e. 1.41% (Fig. 21). The analysis carried out on h leads to Fig. 22.*

Production/delivery lag (lead time)

A key assumption in the EOQ problem formulation is that orders are delivered immediately. We will now relax this assumption and assume that orders are subject to a production/delivery lag of L periods.

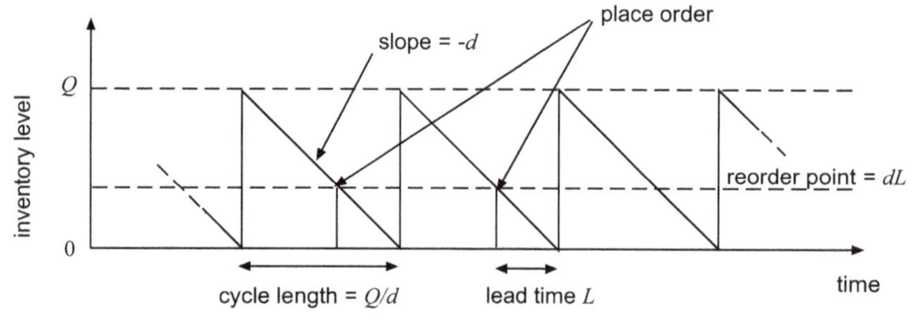

Fig. 23 EOQ inventory curve under production/delivery lag.

By observing the behaviour of the inventory curve in Fig. 23 it is easy to see that the optimal solution does not change. The only adjustment required is to place an order L periods before the inventory level reaches zero. To determine when it is time to issue an order it is convenient to introduce the following definition.

Definition 3 (Reorder point)**.** *The reorder point r is the amount of demand observed over the lead time L*

$$r = dL.$$

Example 6. *In the numerical example presented in Listing 19, assuming $L = 0.5$, the reorder point is 5, which means an order is issued as soon as inventory drops to 5 units. The behaviour of inventory over time is shown in Fig. 24.*

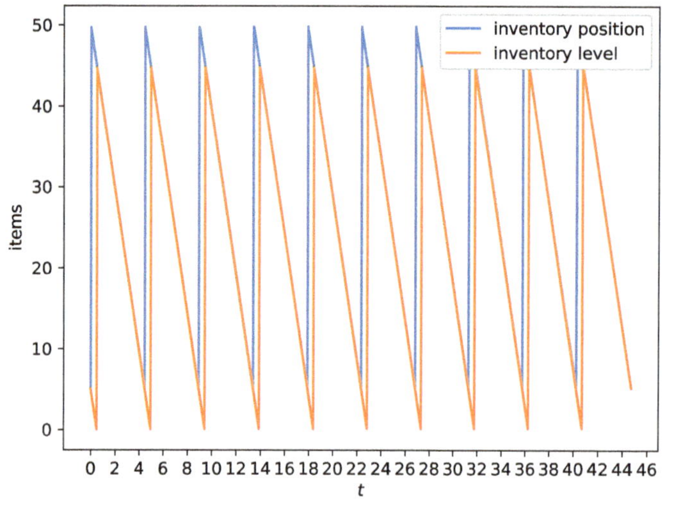

Fig. 24 Behaviour of inventory over time for the numerical example presented in Listing 19, assuming $L = 0.5$.

Powers-of-two policies

The problem statement resembles an EOQ setting; however rather than choosing an arbitrary optimal cycle length T, we are given a base planning period T_b and we must choose an optimal cycle length taking the form $T_b 2^k$, where $k \in \{0, \ldots, \infty\}$. This is particularly useful in settings in which we seek order coordination across a range of stock keeping units.

Recall that $Q^* = dT^*$, where T^* denotes the optimal cycle length of the EOQ. By substituting in $C_r(Q)$ we can express the relevant cost as a function $F(T)$ of the replenishment cycle length T

$$F(T) = \frac{K}{T} + \frac{hdT}{2}.$$

Lemma 8 (Powers-of-two policy). *Let $T_b 2^k$ be a powers-of-two policy with base planning period T_b, the optimal k is the smallest integer k satisfying*

$$F(T_b 2^k) \leq F(T_b 2^{k+1}).$$

Proof. From Lemma 4 it immediately follows that $F(T)$ is convex. □

Lemma 9 (Powers-of-two bound). *Let T_b be a base planning period, then*

$$\frac{F(T_b 2^k)}{F(T^*)} \leq \frac{3}{2\sqrt{2}} \approx 1.06.$$

Proof. From Eq. 4 we obtain

$$\begin{aligned}\frac{F(T)}{F(T^*)} &= \frac{1}{2}\left(\frac{T^*}{T} + \frac{T}{T^*}\right)\\ &= \frac{1}{2}e\left(\frac{T^*}{T}\right)\end{aligned}$$

where $e(x) = x + 1/x$. Since

$$F(T_b 2^k) \leq F(T_b 2^{k+1}) \rightarrow \sqrt{2}^{-1} T^* \leq T_b 2^k = T$$

$$F(T_b 2^{k-1}) > F(T_b 2^k) \rightarrow T = T_b 2^k \leq \sqrt{2} T^*,$$

therefore $\frac{1}{\sqrt{2}} \leq \frac{T^*}{T} \leq \sqrt{2}$ and

$$\frac{F(T_b 2^k)}{F(T^*)} \leq \frac{1}{2}e\left(\frac{1}{\sqrt{2}}\right) = \frac{1}{2}e(\sqrt{2}) = \frac{3}{2\sqrt{2}} \approx 1.06.$$

□

An optimal powers-of-two policy can be computed as shown in Listing 24.

```
class eoq:
    def opt_powersoftwo_policy(self, T:
        float) -> float:
      K, d, h = self.K, self.d, self.h
      rc = lambda t : K/t + h*d*t/2
      k = 0
      while rc(T*2**(k+1)) <
          rc(T*2**k):
        k += 1
      return T*2**k
```

Listing 24 Computing an optimal powers-of-two policy.

Example 7. *In the numerical example presented in Listing 19, given a base planning period $T_b = 0.7$, the ratio $F(T_b 2^k)/F(T^*) = 1.025 \leq 1.06$; hence the resulting powers-of-two policy is only 2.5% more expensive than the optimal one.*

Quantity discounts

In several practical situations it is common to offer a discount on the purchasing price when the order quantity is high enough. In this problem setting we are given breakpoints $b_0, \ldots, b_{T+1}$, where $b_0 = 0$ and $b_{T+1} = \infty$, and associated purchasing prices v_k for $k = 1, \ldots, T+1$, where purchasing price v_k applies within the range (b_{k-1}, b_k). The structure of an instance is illustrated in Listing 25.

Listing 25 EOQ under quantity discounts.

```
class eoq_discounts(eoq):
    def __init__(self, K: float, h: float, d: float, b: List[float], v: List[float]):
        """
        Constructs an instance of the Economic Order Quantity problem.

        Arguments:
            K {float} -- the fixed ordering cost
            h {float} -- the proportional holding cost as a percentage of purchase cost
            d {float} -- the demand per period
            b {float} -- a list of puchasing cost breakpoints
            v {float} -- a list of decreasing unit purchasing costs where v[j] applies
                in (b[j],b[j-1])
        """

        self.K, self.h, self.d, self.b, self.v = K, h, d, b, v
        self.b.insert(0, 0)
        self.b.append(float("inf"))

    def compute_eoq(self) -> float:
        """
        Computes the Economic Order Quantity.

        Returns:
            float -- the Economic Order Quantity
        """

        quantities = [minimize(self.cost,
                               self.b[j-1]+1,
                               bounds=((self.b[j-1],self.b[j]),),
                               method='SLSQP',
                               options={'ftol': 1e-8, 'disp': False}).x[0]
                      for j in range(1, len(self.b))]
        costs = [self.cost(k) for k in quantities]
        return quantities[costs.index(min(costs))]
```

We still observe the fixed ordering cost K and, as discussed, item purchasing price takes different values depending on the size of the order. Holding cost h, however, is no longer absolute and now denotes a percentage of the purchasing price.

Listing 25 also embeds a method `compute_eoq` which computes the economic order quantity. The total cost function is convex within each interval (b_{k-1}, b_k); `compute_eoq` analyses each individual interval (b_{k-1}, b_k) separately and returns the optimal Q that minimizes the total cost across all possible intervals. This solution method works for any possible discount structure.

There are two types of discount strategies typically applied: all-units discounts and incremental discounts.

In **all-units discounts** purchasing price v_k applies to the entire order quantity if this falls within the range (b_{k-1}, b_k).

In **incremental discounts** purchasing price v_k only applies to the fraction of order quantity that falls within within the range (b_{k-1}, b_k).

All-units discounts

Variable ordering cost as a function of the ordering quantity Q (`unit_cost`), as well as variable ordering cost (`co_variable`) and inventory holding cost (`ch`) per replenishment cycle can be computed as shown in Listing 26.

In this case, assuming $Q \in [b_j, b_{j+1})$, the total cost takes the form

$$C(Q) = \frac{Kd}{Q} + \frac{hv_jQ}{2} + v_jd$$

Example 8. *We consider the numerical example in Listing 19. However, we consider all-units discounts with $b = \{0, 10, 20, 30, \infty\}$ and associated $v = \{5, 4, 3, 2\}$. The per unit purchasing cost as a function of the order quantity Q is shown in Fig. 25. The total cost function is shown in Fig. 26. The economic order quantity is $Q^* = 31.6$ and the total cost is $C(Q^*) = 83.2$.*

```
class eoq_all_units(eoq_discounts):
    def unit_cost(self, Q):
        j = set(filter(lambda j:
                self.b[j-1] <= Q <
                self.b[j],
                range(1,len(self.b)))).pop()
        return self.v[j-1]*Q

    def co_variable(self, Q):
        j = set(filter(lambda j:
                self.b[j-1] <= Q <
                self.b[j],
                range(1,len(self.b)))).pop()
        return self.v[j-1]*self.d

    def ch(self, Q: float) -> float:
        j = set(filter(lambda j:
                self.b[j-1] <= Q <
                self.b[j],
                range(1,len(self.b)))).pop()
        h = self.h*self.v[j-1]
        return h*Q/2
```

Listing 26 EOQ under all units quantity discounts.

Fig. 25 All units quantity discounts.

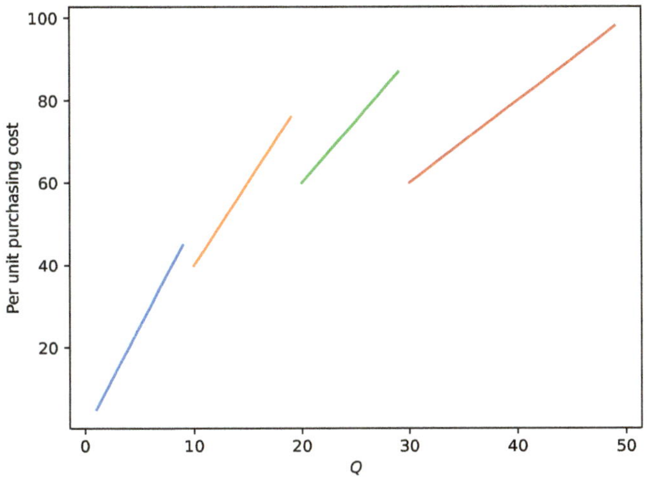

Fig. 26 EOQ total cost for all units quantity discounts.

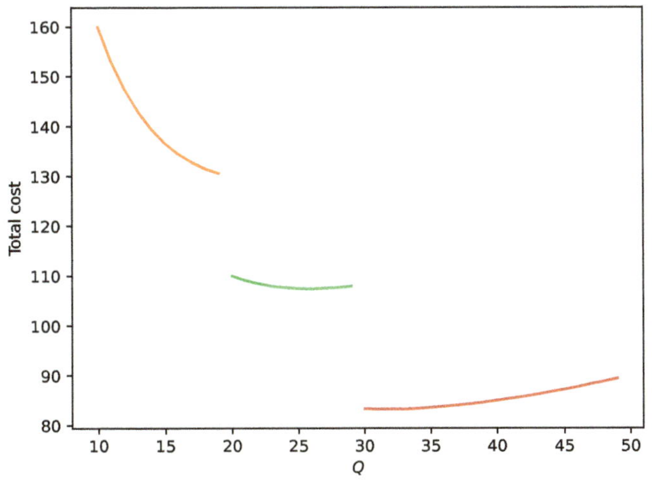

Incremental discounts

Variable ordering cost as a function of the ordering quantity Q (`unit_cost`), as well as variable ordering cost (`co_variable`) and inventory holding cost (`ch`) per replenishment cycle can be computed as shown in Listing 27.

In this case, the total cost takes the form

$$C(Q) = \frac{Kd}{Q} + \frac{h\frac{c(Q)}{Q}Q}{2} + \frac{c(Q)}{Q}d$$

where, assuming $Q \in [b_j, b_{j+1})$,

$$c(Q) = \sum_{i=0}^{j-1} v_i(b_{i+1} - b_i) + v_j(Q - b_j).$$

Example 9. *We consider the numerical example in Listing 19. However, we consider all-units discounts with $b = \{0, 10, 20, 30, \infty\}$ and associated $v = \{5, 4, 3, 2\}$. The per unit purchasing cost as a function of the order quantity Q is shown in Fig. 27. The total cost function is shown in Fig. 28. The economic order quantity is $Q^* = 39.9$ and the total cost is $C(Q^*) = 130$.*

```
class eoq_incremental(eoq_discounts):
    def unit_cost(self, Q):
        j = set(filter(lambda j:
                self.b[j-1] <= Q <
                self.b[j],
                range(1,len(self.b)))).pop()
        return sum([(self.b[k] -
                self.b[k-1]) * self.v[k-1]
                for k in range(1,j)]) + (Q
                - self.b[j-1]) *
                self.v[j-1]

    def co_variable(self, Q):
        j = set(filter(lambda j:
                self.b[j-1] <= Q <
                self.b[j],
                range(1,len(self.b)))).pop()
        cQ = sum([(self.b[k] -
                self.b[k-1]) * self.v[k-1]
                for k in range(1,j)]) + (Q
                - self.b[j-1]) *
                self.v[j-1]
        return self.d*cQ/Q

    def ch(self, Q: float) -> float:
        j = set(filter(lambda j:
                self.b[j-1] <= Q <
                self.b[j],
                range(1,len(self.b)))).pop()
        cQ = sum([(self.b[k] -
                self.b[k-1]) * self.v[k-1]
                for k in range(1,j)]) + (Q
                - self.b[j-1]) *
                self.v[j-1]
        h = self.h*cQ/Q
        return h*Q/2
```

Listing 27 EOQ under incremental quantity discounts.

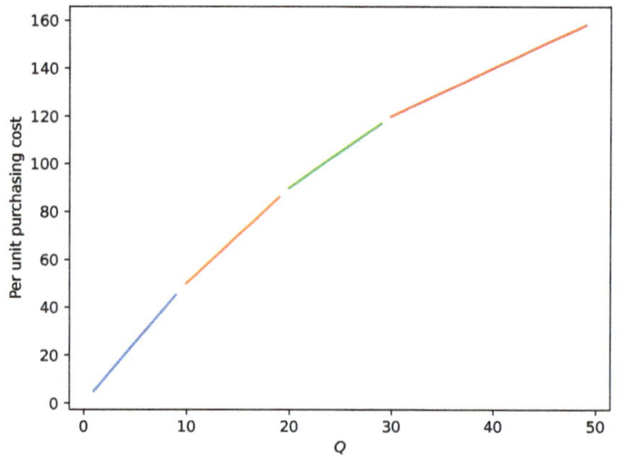

Fig. 27 Incremental quantity discounts.

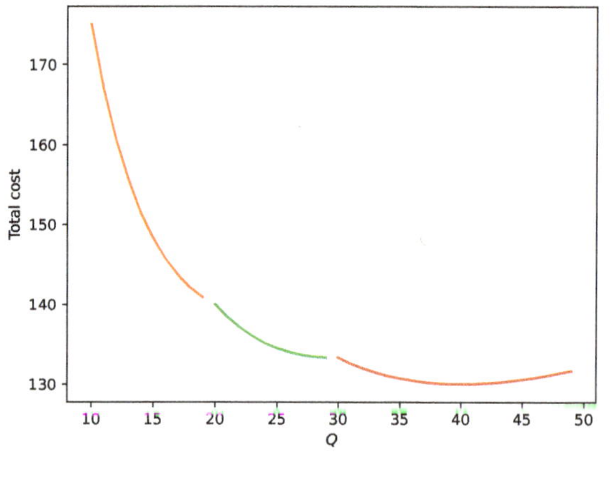

Fig. 28 EOQ total cost for incremental quantity discounts.

Planned backorders in the EOQ

In this section we still consider an EOQ setting, but we relax the assumption that all demand must be met on time. In other words, we will allow demand to be backordered and met when the successive replenishment arrives. The behaviour of the system is illustrated in Fig. 29. The system resembles the classical EOQ. However, the zero-inventory ordering property does not hold for this system. Instead, an order will be issued when inventory reaches $-S$, the planned backorder level. We therefore now have two decision to be made: how much to order (Q) and how much to backorder (S).

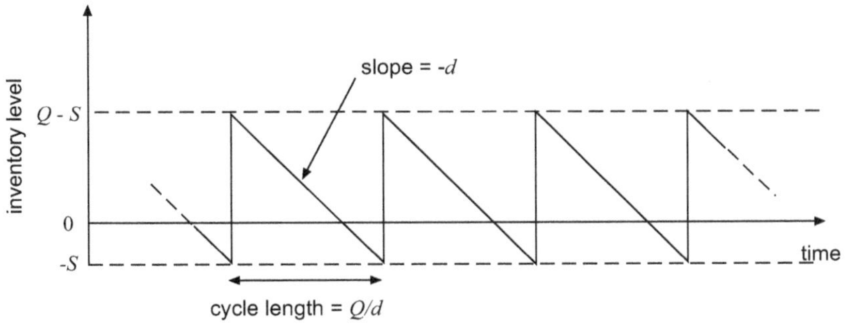

Fig. 29 EOQ inventory curve under planned backorders.

Cost analysis

Incurring backorders must be expensive, otherwise the optimal policy would simply be to not order at all. More specifically, we will charge a penalty cost p per unit backordered per period. The total relevant cost then becomes

$$C_r^b(Q,S) = \underbrace{\frac{Kd}{Q}}_{\text{ordering}} + \underbrace{h\frac{(Q-S)^2}{2Q}}_{\text{holding}} + \underbrace{p\frac{S^2}{2Q}}_{\text{penalty}}$$

Lemma 10 (Holding cost reduction factor). *Allowing backorders is mathematically equivalent to reducing the holding cost rate by the factor* $p/(p+h)$.

Proof. To prove this, it is convenient to let $S = xQ$, where x denotes the fraction of backordered demand in a replenishment cycle. Substitute in $C_r^b(Q,S)$, take partial derivatives w.r.t. Q and x, and set both partial derivatives to zero. Interestingly,

$$\frac{\mathrm{d}C_r^b(Q,S)}{\mathrm{d}x} = -hQ(1-x) + pQx = 0$$

admits solution

$$x^* = \frac{h}{p+h} \tag{5}$$

which is independent of Q. If we plug x^* into $C_r^b(Q,S)$, where $S = xQ$, we then obtain

$$C_r^b(Q) = \frac{hp}{h+p}\frac{Q}{2} + \frac{Kd}{Q} \tag{6}$$

which is the EOQ cost function in which the holding cost rate has been reduced by the factor $p/(p+h)$. □

The planned backorder cost analysis can be implemented as shown in Listing 28.

Listing 28 Planned backorder cost analysis.

```
class eoq_planned_backorders:
    def __init__(self, K: float, h: float, d: float, v: float, p: float):
        """
        Constructs an instance of the Economic Order Quantity problem.

        Arguments:
            K {float} -- the fixed ordering cost
            h {float} -- the proportional holding cost
            d {float} -- the demand per period
            v {float} -- the unit purchasing cost
            p {float} -- the backordering penalty cost
        """
        self.K, self.h, self.d, self.v, self.p = K, h, d, v, p

    def relevant_cost(self, Q: float) -> float:
        return self.co_fixed(Q)+self.ch(Q)+self.cp(Q)

    def cost(self, Q: float) -> float:
        return self.co_fixed(Q)+self.co_variable(Q)+self.ch(Q)+self.cp(Q)

    def co_fixed(self, Q: float) -> float:
        K, d= self.K, self.d
        return K/(Q/d)

    def co_variable(self, Q: float) -> float:
        d, v = self.d, self.v
        return d*v

    def ch(self, Q: float) -> float:
        h = self.h
        x = self.h/(self.p+self.h)
        return h*(Q-Q*x)**2/(2*Q)

    def cp(self, Q: float) -> float:
        p = self.p
        x = self.h/(self.p+self.h)
        return p*(Q*x)**2/(2*Q)
```

Optimal solution

We next characterize the structure of the optimal solution by building upon Lemma 10.

Lemma 11 (Optimal order quantity). *The optimal order quantity is*

$$Q^* = \sqrt{\frac{2Kd(h+p)}{hp}} \tag{7}$$

Lemma 12 (Optimal fraction of backordered demand). *The optimal fraction of backordered demand in a replenishment cycle $x^* = h/(p+h)$.*

Lemma 13 (Optimal cost).

$$C_r^b(Q^*, x^*) = \sqrt{2Kdhp/(h+p)} \tag{8}$$

The computation is similar to that presented in Listing 18.

Example 10. *In the numerical example presented in Listing 19 we consider a penalty cost $p = 5$; then $Q^* = 48.99$, $x^* = 0.16$, $C_r^b(Q^*, x^*) = 40.82$.*

Finite production rate: The Economic Production Quantity

In contrast to the previous section, we shall now relax the assumption that the whole replenishment quantity Q is delivered at once at the beginning of the planning horizon. This problem is known as the Economic Production Quantity (EPQ).[14] The quantity is instead delivered at a constant and finite production rate $p > d$, where d is the demand rate. Taft's original drawings are shown in Fig. 30. The behaviour of the system is illustrated in Fig. 31.

[14] E. W. Taft. The most economical production lot. *The Iron Age*, 101: 1410–1412, 1918.

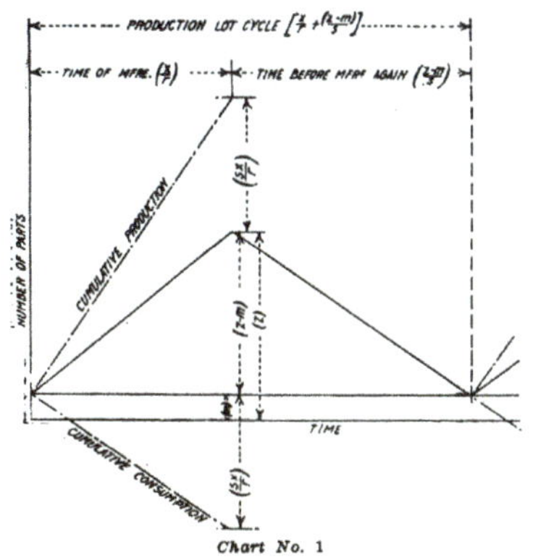

Fig. 30 Taft's inventory curve from [Taft, 1918] (courtesy of HathiTrust).

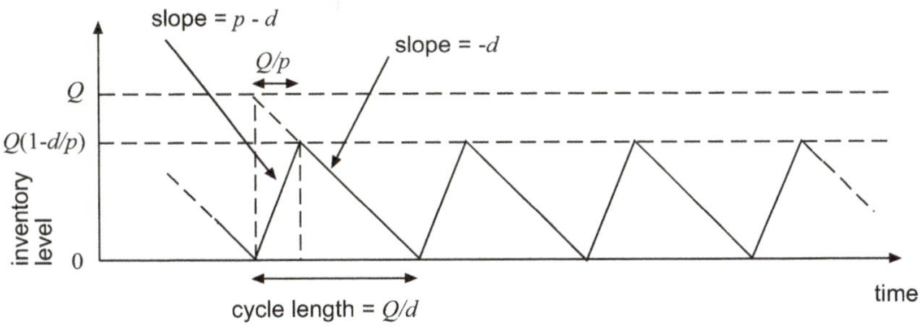

Fig. 31 EPQ inventory curve.

Like in the classical EOQ, the zero inventory ordering property holds for this system. A replenishment occurs when inventory level is zero. Production runs until the whole replenishment quantity Q is delivered. While the replenishment quantity is delivered, demand occurs at rate d, so inventory increases at a rate $p - d$ for Q/p time periods until it reaches a maximum level $Q(1 - d/p)$, and then decreases at rate d over the rest of the replenishment cycle.

Cost analysis

By observing that the average inventory level is now $Q(1 - p/d)/2$ we obtain the following expression for the total relevant cost

$$C_r^p(Q) = \frac{dK}{Q} + \frac{hQ(1-p/d)}{2}. \tag{9}$$

The EPQ analysis can be computed as shown in Listing 29.

```
class epq:
    def __init__(self, K: float, h:
            float, d: float, v: float, p:
            float):
        """
        Constructs an instance of the
            Economic Production
            Quantity problem.

        Arguments:
          K {float} -- the fixed
              ordering cost
          h {float} -- the proportional
              holding cost
          d {float} -- the demand per
              period
          v {float} -- the unit
              purchasing cost
          p {float} -- the finite
              production rate
        """
        self.K, self.h, self.d, self.v,
            self.p = K, h, d, v, p

    def relevant_cost(self, Q: float)
        -> float:
        return
            self.co_fixed(Q)+self.ch(Q)

    def cost(self, Q: float) -> float:
        return self.co_fixed(Q) +
            self.co_variable(Q) +
            self.ch(Q)

    def co_fixed(self, Q: float) ->
        float:
        K, d= self.K, self.d
        return K/(Q/d)

    def co_variable(self, Q: float) ->
        float:
        d, v = self.d, self.v
        return d*v

    def ch(self, Q: float) -> float:
        h = self.h
        rho = self.p/self.d
        return h*Q*(1-rho)/2
```

Listing 29 Economic Production Quantity cost analysis.

Optimal solution

As in the previous case, the optimal solution is simply a minor modification of the classical EOQ solution

Lemma 14 (Economic Production Quantity).

$$Q^* = \sqrt{\frac{2Kd}{h(1-d/p)}} \tag{10}$$

Lemma 15 (Optimal cost).

$$C_r^p(Q^*) = \sqrt{2Kdh(1-d/p)}. \tag{11}$$

The computation is similar to that presented in Listing 18.

Example 11. *In the numerical example presented in Listing 19 we consider a production rate $p = 5$; then $Q^* = 63.24$ and $C_r^p(Q^*) = 31.62$.*

Production on a single machine: The Economic Lot Scheduling

Consider an EPQ problem, for the sake of convenience, we shall divide a production cycle into two phases: the "ramp up" phase and the "depletion" phase. As we have seen in the previous section, a replenishment occurs when inventory level is zero. Production runs until the whole replenishment quantity Q is delivered. While the replenishment quantity is delivered, demand occurs at rate d, so inventory increases at a rate $p-d$ for Q/p time periods ("ramp up" phase) until it reaches a maximum level $Q(1-d/p)$; and then decreases at rate d over the rest of the replenishment cycle, for $Q(1-p/d)/d$ time periods ("depletion" phase). Observe that, naturally, the cycle length is the sum of the length of these two phases: $Q(1-d/p)/d+Q/p=Q/d$ (Fig. 32).

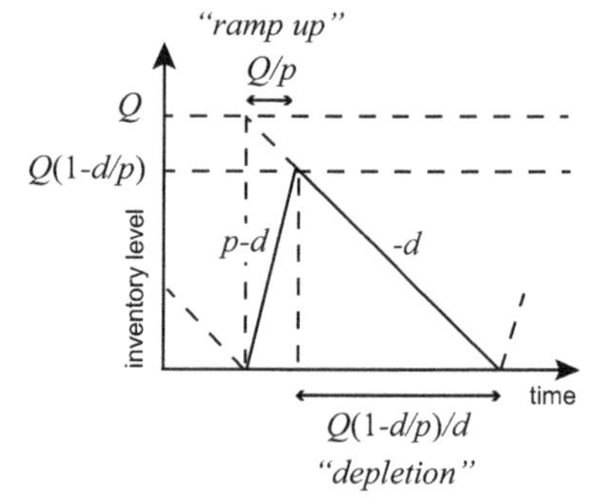

Fig. 32 EPQ inventory curve: "ramp up" and "depletion" phases.

Assume now that the production facility requires a setup time s (e.g. cleaning, maintenance) before a production run may start. Let Q^* be the EPQ (Lemma 14). If $s \leq Q(1-p/d)/d$, the solution is clearly feasible and optimal for the EPQ under setup time, since production facility setup can occur during the "depletion" phase, when the machine is idle, without affecting the "ramp up" phase.

Example 12. *In the numerical example presented in Listing 19, consider a production rate* $p=5$ *and recall that* $d=10$*; then* $Q^*=63.24$*, and* $Q^*(1-p/d)/d=3.162$*. If* $s \leq 3.162$*,* Q^* *remains the EPQ.*

If, however, $s > Q^*(1-p/d)/d$, since the total relevant cost is convex, then the EPQ can be computed by enforcing the condition $s=Q(1-p/d)/d$. In other words, the optimal cycle length turns out to be the cycle length that corresponds to a schedule in which the machine is never idle: it is either producing during the "ramp up" phase, or being setup during the "depletion" phase.

The general expression for the EPQ under setup time is then

Lemma 16 (Economic Production Quantity under setup time).

$$Q^* = \max\left\{\sqrt{\frac{2Kd}{h(1-d/p)}}, ds/(1-p/d)\right\}. \tag{12}$$

Having considered the production of a single item on on a single machine with finite production capacity and setup time, we now generalise our analysis to the production of n items on a single machine with finite production capacity and item dependent setup times.[15]

[15] Jack Rogers. A computational approach to the economic lot scheduling problem. *Management Science*, 4(3): 264–291, 1958.

Let $p_i > d_i$ be the constant and finite production rate of item i, where d_i is the demand rate for item i. If we allow arbitrary production schedules, a feasible solution exists if and only if $\sum_{i=1}^{n} d_i/p_i < 1$; however, finding an optimal production schedule is NP-hard and there is no closed form expression or efficient algorithm readily available. We will therefore focus on determining the best production cycle that contains a single run of each item. This means the cycle lengths of the n items have to be identical. Such a schedule is referred to as a *rotation schedule*.

Rotation Schedule

Finding a rotation schedule when item setup costs K_i are independent across items is no more difficult than solving the single item problem. Let h_i be the holding cost per period for item i and consider a cycle of length T.

Lemma 17. *The average inventory level of item i during a cycle is* $d_i T(1 - d_i/p_i)/2$.

Proof. The length of the production run of item i in a cycle is Td_i/p_i. As we have seen, a production run for item i must start only when inventory of item i is zero. During production (the "ramp up" phase), the level increases at rate $p_i - d_i$, until it reaches level $Td_i(p_i - d_i)/p_i$. After production (the "depletion" phase), inventory decreases at a rate d_i until it reaches zero and a new production run starts. □

Lemma 18. *The total relevant cost per unit time is*

$$C_r^e(T) = \sum_{i=1}^{n} \left(\frac{1}{2} h_i d_i T(1 - d_i/p_i) + \frac{K_i}{T} \right).$$

Proof. Follows from Lemma 17 and from the fact that the average cost per unit time due to setups for item i is K_i/T. □

Lemma 19. *The optimal cycle length is*

$$T^* = \sqrt{\left(\sum_{i=1}^{n} K_i\right) \bigg/ \left(\sum_{i=1}^{n} \frac{h_i d_i (p_i - d_i)}{2p_i}\right)}.$$

Proof. Take the derivative of $C_r^e(T)$ and set it to zero. □

Example 13. *Consider the instance illustrated in Table 2, in Fig. 33 we plot $C_r^e(T)$. The optimal cycle length is $T^* = 1.78$. The rotation schedule is illustrated in Fig. 34; and in Fig. 35 for positive setup times.*

Item	1	2	3
d_j	50	50	60
p_j	400	400	500
h_j	20	20	30
K_j	2000	2500	800

Table 2 Problem parameters for the ELS problem instance.

The ELS analysis can be computed as shown in Listing 30.

If we include setup times s_i that are sequence independent, the problem remains easy since the sum of the setup times will not depend on the production sequence. If the sum of the setup times is less than the idle time in the rotation schedule, the rotation schedule obtained by ignoring setup times remains optimal. Otherwise, as in the single item case, since the total relevant cost is convex, the optimal cycle length can be found by forcing the idle time to be equal to the sum of the setup times (Lemma 20).

```
class els:
  def __init__(self, p: List[float],
        d: List[float], h:
        List[float], s: List[float],
        K: List[float]):
    """
    Constructs an instance of the
        Economic Lot Scheduling
        problem.

    Args:
      p (List[float]): constant
          production rates
      d (List[float]): constant
          demand rates
      h (List[float]): inventory
          holding costs
      s (List[float]): set up times
      K (List[float]): set up costs
    """
    self.p, self.d, self.h, self.s,
        self.K = p, d, h, s, K

  def item_relevant_cost(self, T:
      float, i: int) -> float:
    return (self.h[i] * T *
        self.d[i] *
        (1-self.d[i]/self.p[i]))/2
        + self.K[i]/T

  def relevant_cost(self, T: float):
    return sum(
        [self.item_relevant_cost(T,
        i) for i in range(0,
        self.n)] )

  def compute_els(self):
    K = sum(self.K)
    H = sum([(self.h[i] * self.d[i]
        * (self.p[i] - self.d[i]))
        / (2*self.p[i]) for i in
        range(0, self.n)])
    return math.sqrt(K/H)

  def compute_cycle_length(self,
      T:float, i: int):
    return T*self.d[i]/self.p[i]

  def compute_max_inventory(self,
      T:float, i: int):
    return T*self.d[i]*(self.p[i] -
        self.d[i])/self.p[i]
```

Listing 30 Economic Lot Scheduling cost analysis.

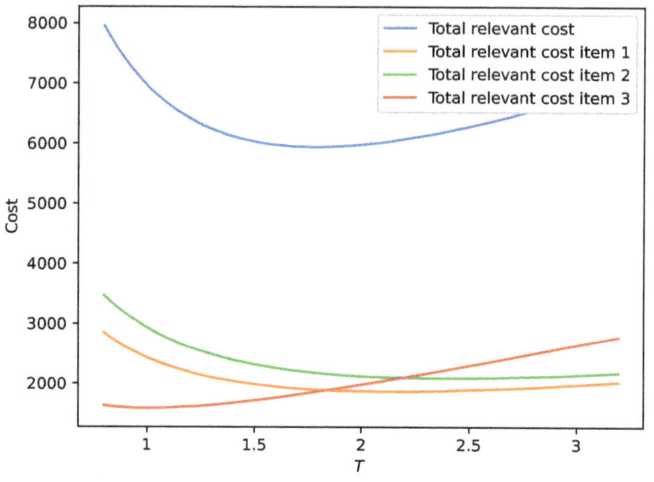

Fig. 33 Overall and item-wise, total relevant cost $C_r^e(T)$ of the ELS problem instance in Example 13.

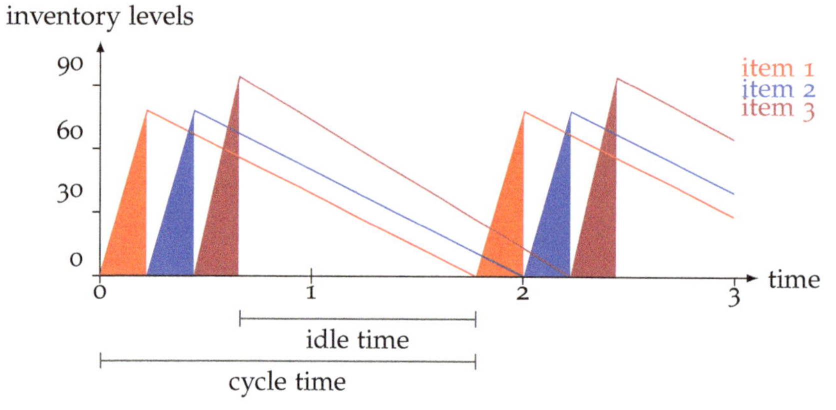

Fig. 34 The optimal rotation schedule for the problem instance in Example 13; solid areas denote production time.

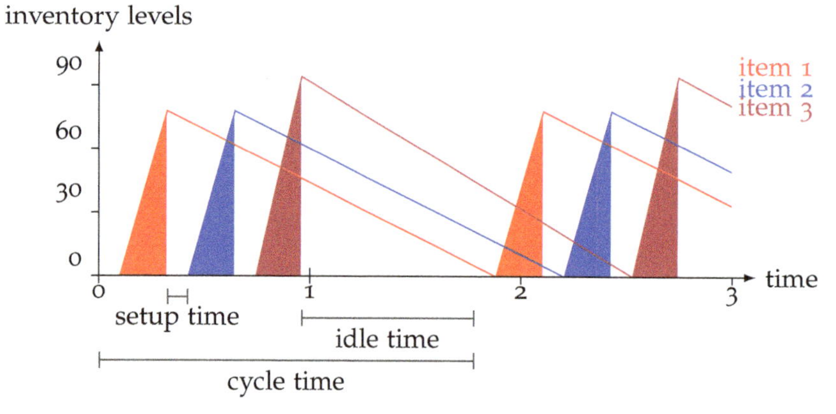

Fig. 35 The optimal rotation schedule for the problem instance in Example 13 assuming all item setup times are equal to 0.1; solid areas denote production time.

Lemma 20. *If the sum of the setup times exceeds the idle time in the rotation schedule, then*

$$T^* = \left(\sum_{i=1}^{n} s_i\right) \Big/ \left(1 - \sum_{i=1}^{n} \frac{d_i}{p_i}\right).$$

Proof. Follows from the convexity of total relevant cost $C_r^e(T)$. □

Synchronising production: The Joint Replenishment Problem

The Joint Replenishment Problem (JRP) occurs when it becomes necessary to synchronise production of multiple items.

Consider a continuous review inventory system comprising n items. Let d_i be the demand rate for item i, and h_i be the holding cost per time period for item i. There are two types of fixed setup costs: the major setup cost K_0 for the system, and a minor setup cost K_i for each item type. Essentially, every time production occurs, the major setup cost K_0 is incurred, regardless of how many types of items are produced. Conversely, the minor setup cost K_i is incurred at time t if and only if item type i is produced at that time. The aim is to minimise the total cost per period.

Two questions must be answered to address the JRP:

- What is the optimal time T_0 between major setups?
- What is the optimal production cycle length T_i for item i?

```python
class jrp:
   def __init__(self, n:int, beta:int,
        h:List[float], d:List[float],
        K:List[float], K0:float):
      """An instance for the Joint
          Replenishment Problem

      Args:
         n (int): the number of items
         beta (int): the base planning
            period
         h (List[float]): holding cost
            rate for item
         d (List[float]): demand rate
            for item
         K (List[float]): fixed minor
            setup cost for item
         K0 (float): fixed major setup
            cost.
      """
      self.n, self.beta, self.h,
           self.d, self.K = n, beta,
           h, d, K
      self.H = [0] + [0.5 * h[i] *
           d[i] for i in range(0,n)]
      self.K = [K0] + K
      self.U = 30 # choose a number
           sufficiently large
```

Listing 31 The JRP in Python.

Lemma 21 (Zero inventory ordering). *It is optimal to produce item i at time t if and only if its inventory level is zero.*

Lemma 22. *The holding cost per time period for item i is* $H_i \triangleq h_i d_i/2$.

The JRP in its general form is an NP-hard problem[16] and, therefore, it is unlikely that an efficient algorithm to solve this problem will be found.

[16] Esther Arkin, Dev Joneja, and Robin Roundy. Computational complexity of uncapacitated multi-echelon production planning problems. *Operations Research Letters*, 8(2):61–66, 1989.

Powers-of-two policies

We shall here focus on a restricted version of the original problem: the JRP under a powers-of-two policy.[17] For each item i, rather than choosing an arbitrary optimal cycle length T_i, we are given a base planning period T_b — which is assumed sufficiently small, and in particular, smaller than the cycle length of the most frequently ordered item — and we must choose an optimal cycle length taking the form $T_b 2^k$, where $k \in \{0,\ldots,\infty\}$. This leads to the following nonlinear programming model (problem Z).

[17] Peter Jackson, William Maxwell, and John Muckstadt. The joint replenishment problem with a powers-of-two restriction. *IIE Transactions*, 17(1): 25–32, 1985.

$$Z : \min \sum_{i=0}^{n} K_i/T_i + H_i T_i \tag{13}$$

Subject to,

$$T_i = M_i T_b \qquad i = 1,\ldots,n, \tag{14}$$
$$T_i \geq T_0 \qquad i = 1,\ldots,n, \tag{15}$$
$$M_i \in \{2^k | k = 0,1,\ldots,\infty\}, \tag{16}$$

where, for the sake of convenience, we let $H_0 \triangleq 0$.

Now, relax constraint 14 in problem Z, and name the new problem obtained $\widehat{Z}$.

Lemma 23. *$\widehat{Z}$ is a lower bound among all feasible policies.*

Lemma 24. *The solution to problem Z is no more than 6% more expensive than the lower bound obtained via problem $\widehat{Z}$.*

Proof. Let T_i^R, $i = 0, \ldots, n$, be the optimal solution to the relaxed problem $\widehat{Z}$. By following a line of reasoning similar to that presented in Lemma 9, one first proves that the powers-of-two restriction implies

$$\frac{1}{\sqrt{2}} \leq \frac{T_i^*}{T_i^R} \leq \sqrt{2}, \tag{17}$$

where T_i^* is the optimal solution to the JRP under a powers-of-two policy (problem Z), $i = 0, \ldots, n$. The result then follows from Eq. 17 and from the convexity of the objective function of $\widehat{Z}$. □

The JRP can be modelled and solved by using ILOG CP Optimizer in Python as shown in Listing 31 and in Listing 32, the solution leverages Constraint Programming[18] to deal with the nonlinear and discrete nature of the problem.

[18] Francesca Rossi, Peter van Beek, and Toby Walsh, editors. *Handbook of Constraint Programming*, volume 2 of *Foundations of Artificial Intelligence*. Elsevier, 2006.

```
from docplex.cp.model import CpoModel
from typing import List
class jrp:
    def solve(self):
        mdl = CpoModel()

        M = mdl.integer_var_list(self.n+1, 0, self.U, "M")
        power = [2**i for i in range(0,self.U+1)]
        T = mdl.integer_var_list(self.n+1, 0, power[self.U], "T")

        mdl.add(mdl.element(power, M[i]) == T[i] for i in range(0,self.n+1))
        mdl.add(T[i] >= T[0] for i in range(0,self.n+1))

        mdl.minimize(mdl.sum(self.H[i]*T[i]/self.beta+self.K[i]/(T[i]/self.beta) for i
            in range(0,self.n+1)))

        print("Solving model....")
        msol = mdl.solve(TimeLimit=10, agent='local', execfile=
          '/Applications/CPLEX_Studio1210/cpoptimizer/bin/x86-64_osx/cpoptimizer')
        msol.print_solution() if msol else print("no solution") # Print solution
```

Listing 32 Solving the JRP by using ILOG CP Optimizer in Python.

```
instance = {"n": 5, "beta": 52,
        "h":[1,1,1,1,1],
        "d":[2,2,2,2,2],
        "K":[1,2,4,6,16], "K0": 5}
jrp = jrp(**instance)
jrp.solve()
```

Fig. 36 A JRP instance; note that beta $= 1/T_b$.

Example 14. *Consider a base planning period $T_b = 1/52$ (e.g. a planning on a weekly basis), and the problem parameters in Table 3. The total cost of a powers-of-two policy is 25.3, the optimal solution is to order items $1, \ldots, 4$ every $2^7 = 128$ weeks, and to order item 5 every 256 weeks. The lower bound obtained by solving the relaxed problem $\widehat{Z}$ is 24.9, then $24.9 + 6\% = 26.4$; as expected, the total cost of a powers-of-two policy falls within the bounds $25.3 \in (24.9, 26.4)$.*

Item	0	1	2	3	4	5
d_i	-	2	2	2	2	2
h_i	0	1	1	1	1	1
K_i	5	1	2	4	6	16

Table 3 Problem parameters for the JRP problem instance (yearly rates).

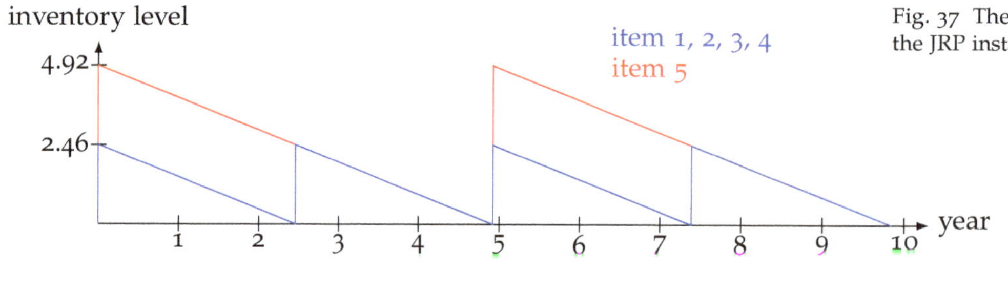

Fig. 37 The optimal ordering plan for the JRP instance in Fig. 3.

Time-varying demand: Dynamic Lot Sizing

While exploring variants of the EOQ problem we maintained the assumption that demand rate is known and constant and that inventory is reviewed continuously. Both these assumptions may result unrealistic in practice. In fact, it is often the case that decision maker operate under a "periodic review" setting in which inventory can be reviewed — and orders issued — only at certain points in time. This leads to a discretization of the planning horizon into periods. Moreover, demand rate in practice is often not constant and varies from period to period.

In their 1958 seminal work[19] Wagner and Whitin explored the Dynamic Version of the Economic Lot Size Model. The so-called Wagner-Whitin problem setting considers a finite planning horizon comprising T periods. Demand d_t may vary from one period t to another. Unlike the EOQ inventory can only be reviewed — an orders issued — at the beginning of each period. Like in the Economic Order Quantity orders are received immediately after being placed, there is a fixed cost K as well as variable cost v for placing an order. There is a proportional cost h for carrying one unit of inventory from one period to the next. Finally, all demand must be met on time and the initial inventory is assumed to be zero. It is safe to disregard the proportional ordering cost because the planning horizon is finite and all demand must be met, therefore this is in fact a constant. The Wagner-Whitin problem can be modelled in Python as shown in Listing 33.

[19] Harvey M. Wagner and Thomson M. Whitin. Dynamic version of the economic lot size model. *Management Science*, 5(1):89–96, 1958.

```
class WagnerWhitin:
    def __init__(self, K: float, h:
        float, d: List[float], I0:
        float):
        """
        Create an instance of a
            Wagner-Whitin problem.

        Arguments:
            K {float} -- the fixed
                ordering cost
            h {float} -- the per unit
                holding cost
            d {List[float]} -- the demand
                in each period
            I0 {float} -- the initial
                inventory level
        """
        self.K, self.h, self.d, self.I0
            = K, h, d, I0
```

Listing 33 The Wagner-Whitin base class.

As in the EOQ, we leverage the concept of replenishment cycle.

Lemma 25. *The cost associated with a replenishment cycle starting in period i and ending in period j (included) is*

$$c(i,j) = K + h\sum_{k=i}^{j}(k-i)d_k.$$

Costs $c(i,j)$ can be computed as shown in Listing 34.

```
class WagnerWhitinDP:
    def cycle_cost(self, i: int, j:
        int) -> float:
        '''
        Compute the cost of a
            replenishment cycle
            covering periods i,...,j
        '''
        if i>j: raise Exception('i>j')

        return self.K + self.h *
            sum([(k-i)*self.d[k] for k
            in range(i,j+1)])
```

Listing 34 Wagner-Whitin cycle cost analysis.

We can then represent the problem as a Directed Acyclic Graph (DAG) in which arcs represent all possible replenishment cycles that can take place within our T-period planning horizon (Fig. 38) and in which the cost associated with arc (i,j) is $c(i,j-1)$.

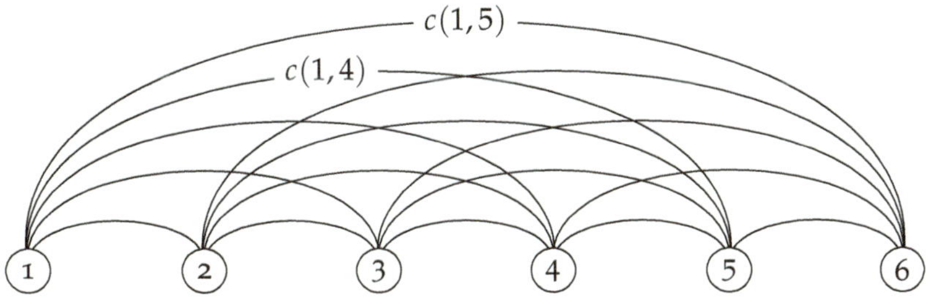

Fig. 38 Wagner-Whitin cost network.

The traditional Wagner-Whitin shortest path algorithm can be implemented in Python as shown in Listing 35.

```
from typing import List
import networkx as nx
import itertools

class WagnerWhitinDP(WagnerWhitin):
    """
    Implements the traditional Wagner-Whitin shortest path algorithm.
    """
    def __init__(self, K: float, h: float, d: List[float]):
        super().__init__(K, h, d, 0)
        self.graph = nx.DiGraph()
        for i in range(0, len(self.d)):
            for j in range(i+1, len(self.d)):
                self.graph.add_edge(i, j, weight=self.cycle_cost(i, j-1))
```

Listing 35 Wagner-Whitin dynamic programming problem setup.

It can be shown that determining the cost of an optimal plan is equivalent to finding the shortest path in the aforementioned DAG. This can be done efficiently, for instance by leveraging Dijkstra's algorithm.[20]

The cost of an optimal plan and associated order quantities can be retrieved as shown in Listing 36.

[20] Edsger W. Dijkstra. A note on two problems in connexion with graphs. *Numerische Mathematik*, 1(1):269–271, 1959.

```
class WagnerWhitinDP:
    def optimal_cost(self) -> float:
        '''
        Compute the cost of an optimal solution to the Wagner-Whitin problem
        '''
        T, cost, g = len(self.d), 0, self.graph
        path = nx.dijkstra_path(g, 0, T-1)
        path.append(len(self.d))
        for t in range(1,len(path)):
            cost += self.cycle_cost(path[t-1],path[t]-1)
        return cost

    def order_quantities(self) -> List[float]:
        '''
        Compute optimal Wagner-Whitin order quantities
        '''
        T, g = len(self.d), self.graph
        path = nx.dijkstra_path(g, 0, T-1)
        path.append(len(self.d))
        qty = [0 for k in range(0,T)]
        for t in range(1,len(path)):
            qty[path[t-1]] = sum([self.d[k] for k in range(path[t-1],path[t])])
        return qty
```

Listing 36 Wagner-Whitin problem solution cost retrieval.

Example 15. *We now consider the Wagner-Whitin problem shown in Listing 37. The cost of an optimal plan is 110, and associated order quantities in each period are* $\{30, 0, 30, 40\}$. *The optimal plan is visualised as a shortest path in Fig. 39.*

```
instance = {"K": 30, "h": 1,
        "d":[10,20,30,40]}
ww = WagnerWhitinDP(**instance)
print("Cost of an optimal plan: ",
        ww.optimal_cost())
print("Optimal order quantities: ",
        ww.order_quantities())
```

Listing 37 A Wagner-Whitin instance.

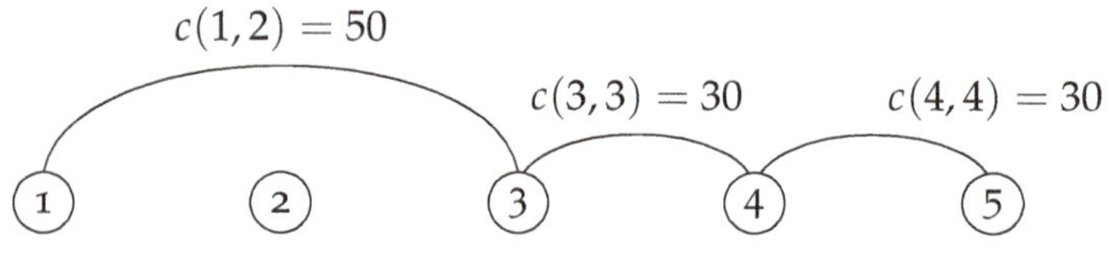

Fig. 39 Wagner-Whitin optimal solution as a shortest path.

Positive initial inventory

Accounting for a positive initial inventory I_0 only requires a small modification to the DAG structure. Essentially, we must compute cycle costs as follows:

$$c(i,j) = \begin{cases} hI_0 + h\sum_{k=i}^{j}(k-i)d_k & \text{if } i=1,\ \sum_{k=1}^{j} d_k \leq I_0; \\ \infty & \text{if } i>1,\ \sum_{k=1}^{j} d_k \leq I_0; \\ K + h\sum_{k=i}^{j}(k-i)d_k & \text{otherwise.} \end{cases}$$

Moreover, while retrieving order quantities, we should bear in mind that we should issue an order in period t, only if $I_0 < \sum_{k=1}^{t} d_k$. Note that if I_0 exceeds the total demand over the planning horizon, then clearly it is optimal to never place any order.

A Wagner-Whitin instance with positive initial inventory is shown in Listing 38. The amended Python code is presented in Listing 39.

```
instance = {"K": 30, "h": 1,
     "d":[10,20,30,40], "I0": 40}
ww = WagnerWhitinDP(**instance)
print("Cost of an optimal plan: ",
     ww.optimal_cost())
print("Optimal order quantities: ",
     ww.order_quantities())
```

Listing 38 A Wagner-Whitin instance.

Listing 39 Wagner-Whitin problem with positive initial inventory.

```
class WagnerWhitinDP(WagnerWhitin):
   """
   Extension of the original Wagner-Whitin algorithm
   to embed a nonnegative initial inventory.
   """

   def __init__(self, K: float, h: float, d: List[float], I0: float):
      super().__init__(K, h, d, I0)
      self.graph = nx.DiGraph()
      for i in range(0, len(self.d)):
         for j in range(i+1, len(self.d)):
            self.graph.add_edge(i, j, weight=self.cycle_cost(i, j-1))

   def cycle_cost(self, i: int, j: int) -> float:
      '''
      Compute the cost of a replenishment cycle covering periods i,...,j
      when initial inventory is nonzero
      '''
      if i>j: raise Exception('i>j')

      if i == 0 and sum(self.d[0:j+1]) <= self.I0:
         return self.h * sum([(k-i)*self.d[k] for k in range(i,j+1)]) + \
               self.h * (j+1) * (self.I0-sum(self.d[0:j+1])) # cost no order
      elif i > 0 and sum(self.d[0:j+1]) <= self.I0:
         return sys.maxsize
      else:
         return self.K + self.h * \
               sum([(k-i)*self.d[k] for k in range(i,j+1)]) # cost with order

   def optimal_cost(self) -> float:
      '''
      Compute the cost of an optimal solution to the Wagner-Whitin problem
      '''
      T, cost, g = len(self.d), 0, self.graph
      path = nx.dijkstra_path(g, 0, T-1)
      path.append(len(self.d))
      for t in range(1,len(path)):
         cost += self.cycle_cost(path[t-1],path[t]-1)
      return cost

   def order_quantities(self) -> List[float]:
      '''
      Compute optimal Wagner-Whitin order quantities
      '''
      T, g = len(self.d), self.graph
      path = nx.dijkstra_path(g, 0, T-1)
      path.append(len(self.d))
      qty = [0 for k in range(0,T)]
      for t in range(1,len(path)):
         qty[path[t-1]] = sum([self.d[k] for k in range(path[t-1],path[t])]) if
               sum(self.d[0:path[t-1]+1]) > self.I0 else 0
      return qty
```

Planned backorders in Dynamic Lot Sizing

We consider an extension of the Wagner-Whitin problem setting in which demand can be backordered from one period to the next.

Consider a finite planning horizon comprising T periods. Demand d_t may vary from one period t to another. Inventory can only be reviewed — an orders issued — at the beginning of each period. Orders are received immediately after being placed, there is a fixed cost K as well as variable cost v for placing an order. There is a proportional cost h for carrying one unit of inventory from one period to the next. There is a proportional backorder/penalty cost p for every unit that is backordered at the end of a period. The initial inventory is assumed to be equal to I_0. This problem can be modelled as follows.

$$\min \quad \sum_{t\in T} \delta_t K + vQ_t + hI_t^+ + pI_t^- \tag{18}$$

Subject to,

$$Q_t \leq M\delta_t \qquad t = 1, \ldots, T \tag{19}$$

$$I_0 + \sum_{k=0}^{t}(Q_k - d_k) = I_t \qquad t = 1, \ldots, T \tag{20}$$

$$I_t = I_t^+ - I_t^- \qquad t = 1, \ldots, T \tag{21}$$

$$Q_t, I_t^+, I_t^- \geq 0 \qquad t = 1, \ldots, T \tag{22}$$

where $M = \sum_{t\in T} d_t$.

The Python code implementing this mathematical programming model is presented in Listing 40 and in Listing 41.

Listing 40 Wagner-Whitin problem with planned backorders, problem instance.

```
# http://ibmdecisionoptimization.github.io/docplex-doc/mp/creating_model.html
# http://www-01.ibm.com/support/docview.wss?uid=swg27042869&aid=1
from docplex.mp.model import Model
import sys
sys.path.insert(0,'/Applications/CPLEX_Studio128/cplex/Python/3.6/x86-64_osx')
from typing import List

class WagnerWhitinPlannedBackorders:
    """
    A Wagner-Whitin problem with planned backorders.

    H.M. Wagner and T. Whitin,
    "Dynamic version of the economic lot size model,"
    Management Science, Vol. 5, pp. 89-96, 1958
    """
    def __init__(self, K: float, v: float, h: float, p: float, d: List[float], I0:
        float):
        """
        Create an instance of a Wagner-Whitin problem.

        Arguments:
            K {float} -- the fixed ordering cost
            v {float} -- the per unit ordering cost
            h {float} -- the per unit holding cost
            p {float} -- the per unit backorder cost
            d {List[float]} -- the demand in each period
            I0 {float} -- the initial inventory level
        """

        self.K, self.v, self.h, self.p, self.d, self.I0 = K, v, h, p, d, I0
```

```
class WagnerWhitinPlannedBackordersCPLEX(WagnerWhitinPlannedBackorders):
    """
    Model and solve the Wagner-Whitin problem as an MILP via CPLEX
    """
    def model(self):
        model = Model("Wagner Whitin planned backorders")

        T, M = len(self.d), sum(self.d)
        idx = [t for t in range(0,T)]

        # Decision variables
        self.Q = model.continuous_var_dict(idx, name="Q")
        I = model.continuous_var_dict(idx, lb=-M, name="I")
        Ip = model.continuous_var_dict(idx, name="I^+")
        Im = model.continuous_var_dict(idx, name="I^-")
        delta = model.binary_var_dict(idx, name="delta")

        # Constraints
        for t in range(0,T):
            model.add_constraint(self.Q[t] <= delta[t]*M) # Eq. 14
            model.add_constraint(self.I0 + model.sum(self.Q[k] - self.d[k] for k in
                 range(0,t+1)) == I[t]) # Eq. 15
            model.add_constraint(I[t] == Ip[t]-Im[t]) # Eq. 16
            model.add_constraint(self.Q[t] >= 0) # Eq. 17a
            model.add_constraint(Ip[t] >= 0) # Eq. 17b
            model.add_constraint(Im[t] >= 0) # Eq. 17c

        model.minimize(model.sum(delta[t] * self.K + self.Q[t] * self.v + self.h *
             Ip[t] + self.p * Im[t] for t in range(0,T))) # Eq. 13

        model.print_information()
        self.msol = model.solve()
        if self.msol:
            model.print_solution()
        else:
            print("Solve status: " + self.msol.get_solve_status() + "\n")

    def order_quantities(self) -> List[float]:
        """
        Compute optimal Wagner-Whitin order quantities
        """
        return [self.msol.get_var_value(self.Q[t]) for t in range(0,len(self.d))]

    def optimal_cost(self) -> float:
        """
        Compute the cost of an optimal solution to the Wagner-Whitin problem
        """
        return self.msol.get_objective_value()
```

Listing 41 Wagner-Whitin problem with planned backorders, cplex model.

The optimal ordering plan for the instance in Listing 42 is illustrated in Fig. 40.

```
instance = {"K": 40, "v": 1, "h": 1,
     "p": 2, "d":[10,20,30,40], "I0":
     0}
p = WagnerWhitinPlannedBackordersCPLEX
     (**instance)
p.model()
```

Listing 42 A Wagner-Whitin with planned backorders problem instance.

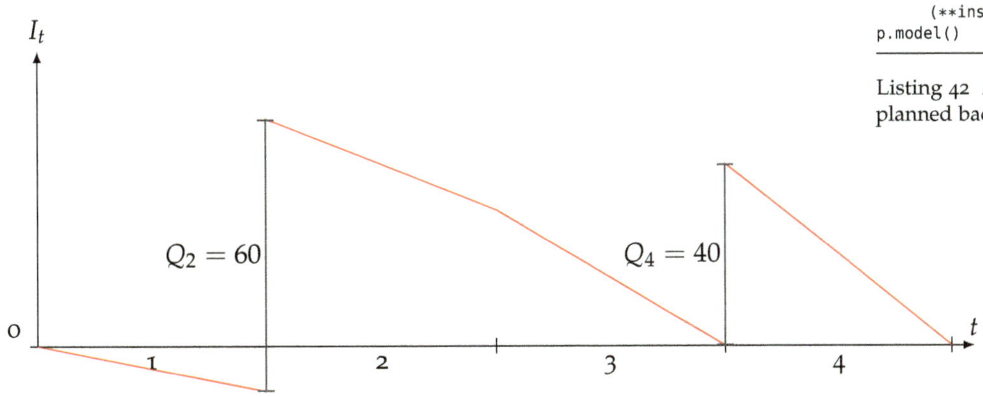

Fig. 40 The optimal ordering plan for the instance in Fig. 42.

Order quantity capacity constraints in Dynamic Lot Sizing

We consider an extension of the Wagner-Whitin problem setting in which capacity constraints are imposed on the order quantity in each period.

Consider a finite planning horizon comprising T periods. Demand d_t may vary from one period t to another. Inventory can only be reviewed — an orders issued — at the beginning of each period. The maximum order quantity in each period is C. Orders are received immediately after being placed. There is a fixed cost K as well as variable cost v for placing an order. There is a proportional cost h for carrying one unit of inventory from one period to the next. Finally, all demand must be met on time and the initial inventory is assumed to be equal to I_0. This problem can be modelled as follows.

$$\min \quad \sum_{t \in T} \delta_t K + vQ_t + hI_t \qquad (23)$$

Subject to,

$$Q_t \leq C\delta_t \qquad t = 1, \ldots, T \qquad (24)$$

$$I_0 + \sum_{k=0}^{t} (Q_k - d_k) = I_t \qquad t = 1, \ldots, T \qquad (25)$$

$$Q_t, I_t \geq 0 \qquad t = 1, \ldots, T \qquad (26)$$

The Python code implementing this mathematical programming model is presented in Listing 43 and in Listing 44.

Listing 43 Capacitated stochastic lot sizing, problem instance.

```
from typing import List

class CapacitatedLotSizing:
    """
    A capacitated lot sizing problem under capacity constraints.

    M. Florian, J. K. Lenstra, and A. H. G. Rinnooy Kan.
    Deterministic production planning: Algorithms and complexity.
    Management Science, 26(7): 669-679, July 1980
    """
    def __init__(self, K: float, v: float, h: float, d: List[float], I0: float, C:
        float):
        """
        Create an instance of the capacitated lot sizing problem.

        Arguments:
            K {float} -- the fixed ordering cost
            v {float} -- the per unit ordering cost
            h {float} -- the per unit holding cost
            d {List[float]} -- the demand in each period
            I0 {float} -- the initial inventory level
            C {float} -- the order capacity
        """

        self.K, self.v, self.h, self.d, self.I0, self.C = K, v, h, d, I0, C
```

The optimal ordering plan for the instance in Listing 45 is illustrated in Fig. 41.

Listing 44 Capacitated lot sizing, cplex model.

```
# http://ibmdecisionoptimization.github.io/docplex-doc/mp/creating_model.html
# http://www-01.ibm.com/support/docview.wss?uid=swg27042869&aid=1
from docplex.mp.model import Model
import sys
sys.path.insert(0,'/Applications/CPLEX_Studio128/cplex/Python/3.6/x86-64_osx')

class CapacitatedLotSizingCPLEX(CapacitatedLotSizing):
    """
    Solves the capacitated lot sizing problem as an MILP.
    """

    def __init__(self, K: float, v: float, h: float, d: List[float], I0, C: float):
        """
        Create an instance of the capacitated lot sizing problem.

        Arguments:
            K {float} -- the fixed ordering cost
            v {float} -- the per unit ordering cost
            h {float} -- the per unit holding cost
            d {List[float]} -- the demand in each period
            I0 {float} -- the initial inventory level
        """
        super().__init__(K, v, h, d, I0, C)
        self.model()

    def model(self):
        """
        Model and solve the capacitated lot sizing problem via CPLEX
        """

        model = Model("Capacitated lot sizing")
        T = len(self.d)
        idx = [t for t in range(0,T)]
        self.Q = model.continuous_var_dict(idx, name="Q")
        I = model.continuous_var_dict(idx, lb=0, name="I")
        delta = model.binary_var_dict(idx, name="delta")


        for t in range(0,T):
            model.add_constraint(self.Q[t] <= delta[t]*self.C)
            model.add_constraint(self.I0 + model.sum(self.Q[k] - self.d[k] for k in
                  range(0,t+1)) == I[t])
            model.add_constraint(self.Q[t] >= 0)
            model.add_constraint(I[t] >= 0)

        model.minimize(model.sum(delta[t] * self.K + self.Q[t] * self.v + self.h *
             I[t] for t in range(0,T)))
        model.print_information()
        self.msol = model.solve()
        if self.msol:
            model.print_solution()
        else:
            print("Solve status: " + self.msol.get_solve_status() + "\n")

    def order_quantities(self) -> List[float]:
        '''
        Compute optimal capacitated lot sizing order quantities
        '''
        return [self.msol.get_var_value(self.Q[t]) for t in range(0,len(self.d))]

    def optimal_cost(self) -> float:
        '''
        Compute the cost of an optimal solution to the capacitated lot sizing problem
        '''
        return self.msol.get_objective_value()
```

Listing 45 Capacitated lot sizing problem instance.

```
instance = {"K": 40, "v": 1, "h": 1, "d":[10,20,30,40], "I0": 0, "C": 30}
CapacitatedLotSizingCPLEX(**instance)
```

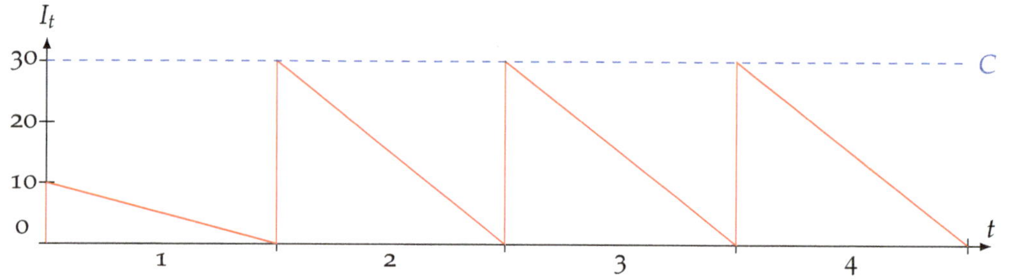

Fig. 41 The optimal ordering plan for the instance in Listing 45.

Computational complexity

The capacitated lot sizing problem is known to be NP-hard.[21] This means that it is unlikely we will ever find an efficient solution method to compute optimal replenishment plans. Apart from the mathematical programming model presented in the previous section, the problem can also be solved via dynamic programming.

[21] Michael Florian, Jan K. Lenstra, and Alexander H. G. Rinnooy Kan. Deterministic production planning: Algorithms and complexity. *Management Science*, 26(7):669–679, 1980.

Dynamic programming formulation

Consider the capacitated lot sizing problem,

- T is the number of periods;
- the state s is the initial inventory level in period t; therefore $S_t \triangleq \{0, \ldots, D\}$, for $t = 1, \ldots, T$, where $D = \sum_{t=1}^{T} d_t$;
- the action a is the order quantity in period t; therefore $A_s \triangleq \{0, \ldots, C\}$ for any state s;
- the state transition function is simply $g_t(s, a) \triangleq s + a - d_t$;
- the immediate cost if action $a \in A_s$ is taken in state $s \in S_t$, is

$$c(s,a) \triangleq \begin{cases} K + av + h\max(s + a - d_t, 0) + M\max(d_t - s + a, 0) & a > 0, \\ h\max(s + a - d_t, 0) + M\max(d_t - s + a, 0) & \text{otherwise,} \end{cases}$$

  where M is a large number;
- the functional equation is

$$f_t(s) = \min_{a \in A_s} c(s,a) + f_{t+1}(g_t(s,a)) \qquad (27)$$

  for which the boundary condition is $f_{T+1}(s) \triangleq 0$ for all $s \in S_{T+1}$.

The goal is to determine $f_t(s)$, where s denotes the initial inventory level in the first period.

WE NEXT SHOW HOW TO MODEL AND SOLVE the capacitated lot sizing problem via dynamic programming[22] in Python.

The `State` class (Listing 46), is used to capture a state of the system. Note that we implement function `__eq__` to ensure to states can be compared with each other.

[22] For more information on dynamic programming refer to the Appendix.

Listing 46 Capacitated lot sizing, auxiliary classes.

```
from typing import List

class State:
    """
    The state of the inventory system.
    """

    def __init__(self, t: int, I: float):
        """Instantiate a state

        Arguments:
            t {int} -- the time period
            I {float} -- the initial inventory
        """

        self.t, self.I = t, I

    def __eq__(self, other):
        return self.__dict__ == other.__dict__

    def __str__(self):
        return str(self.t) + " " + str(self.I)

    def __hash__(self):
        return hash(str(self))
```

Listing 47 Capacitated lot sizing, stochastic dynamic programming model (part 1 of 2).

```
class CapacitatedLotSizingSDP(CapacitatedLotSizing):
    """
    Solves the capacitated lot sizing problem as an SDP.
    """

    def __init__(self, K: float, v: float, h: float, d: List[float], I0, C: float):
        """
        Create an instance of the capacitated lot sizing problem.

        Arguments:
            K {float} -- the fixed ordering cost
            v {float} -- the per unit ordering cost
            h {float} -- the per unit holding cost
            d {List[float]} -- the demand in each period
            I0 {float} -- the initial inventory level
        """
        super().__init__(K, v, h, d, I0, C)

        # initialize instance variables
        self.T, min_inv, max_inv, M = len(d), 0, sum(d), 100000

        # lambdas
        self.ag = lambda s: [x for x in range(0, min(max_inv-s.I, self.C+1))] # action
             generator
        self.st = lambda s, a, d: State(s.t+1, s.I+a-d) # state transition
        L = lambda i,a,d : self.h*max(i+a-d, 0) + M*max(d-i-a, 0) # immediate
             holding/penalty cost
        self.iv = lambda s, a, d: (self.K+v*a if a > 0 else 0) + L(s.I, a, d) #
             immediate value function

        self.cache_actions = {} # cache with optimal state/action pairs

        print("Total cost: " + str(self.f(self.I0)))
        print("Order quantities: " + str([Q for Q in self.order_quantities()]))

    def _compute_order_quantities(self):
        '''
        Compute optimal capacitated lot sizing order quantities
        '''
        I = self.I0
        for t in range(len(self.d)):
            Q = self.q(t, I)
            I += Q - self.d[t]
            yield Q

    def order_quantities(self) -> List[float]:
        return [Q for Q in self._compute_order_quantities()]

    def optimal_cost(self) -> float:
        '''
        Compute the cost of an optimal solution to the capacitated lot sizing problem
        '''
        return self.f(self.I0)
```

Listing 48 Capacitated lot sizing, stochastic dynamic programming model (part 2 of 2).

```
class CapacitatedLotSizingSDP(CapacitatedLotSizing):
    """
    Solves the capacitated lot sizing problem as an SDP.
    [...continues...]
    """

    @memoize
    def _f(self, s: State) -> float:
        """
        Dynamic programming forward recursion.

        Arguments:
            s {State} -- the initial state

        Returns:
            float -- the cost of an optimal policy
        """
        #Forward recursion
        v = min( # optimal cost
            [(self.iv(s, a, self.d[s.t])+ # immediate cost
             (self._f(self.st(s, a, self.d[s.t])) if s.t < self.T-1 else 0)) # future
                 cost
             for a in self.ag(s)]) # actions

        opt_a = lambda a: (self.iv(s, a, self.d[s.t])+ # optimal cost
                  (self._f(self.st(s, a, self.d[s.t])) if s.t < self.T-1 else 0)) == v

        q = [k for k in filter(opt_a, self.ag(s))] # retrieve best action list
        self.cache_actions[str(s)]=q[0] if bool(q) else None # store an action in
             dictionary
        return v # return expected total cost

    def f(self, level: float) -> float:
        """
        Recursively solve the capacitated lot sizing problem
        for an initial inventory level.

        Arguments:
            level {float} -- the initial inventory level

        Returns:
            float -- the cost of an optimal policy
        """

        s = State(0,level)
        return self._f(s)

    def q(self, period: int, level:float) -> float:
        """
        Retrieves the optimal order quantity for a given initial inventory level.

        Arguments:
            period {int} -- the initial period
            level {float} -- the initial inventory level

        Returns:
            float -- the optimal order quantity
        """

        s = State(period,level)
        if not(str(s) in self.cache_actions):
            self._f(s)
        return self.cache_actions[str(s)]
```

To model this problem, we adopted a trick: it is clear that, since inventory cannot be negative, given a period t and an initial inventory level, some actions may be infeasible, therefore in general the space of possible action A_s may not be equal to $\{0, \ldots, C\}$. It may be complex to determine what values $a \in \{0, \ldots, C\}$ are feasible for a given state s. To overcome this difficulty, we allowed potentially infeasible actions, but we associated a very high cost (the large number M) to infeasible states in the immediate cost function $c(s, a)$.

IN DYNAMIC PROGRAMMING, an optimal solution can be obtained via forward recursion or backward recursion. In Listing 47 and Listing 48 we present a solution based on forward recursion.

The action generator function, the state transition function, and the immediate value function are conveniently captured in Python via `lambda` expressions. A generic dynamic programming forward recursion that leverages these lambda expressions is implemented in function `_f`; this function is a direct implementation of Eq. 27.

Finally, the `memoize` class (Listing 49) is a decorator[23] used to tabulate function `_f` and make sure that, if $f_t(s)$ has been already computed for a given state s, this computation does not happen twice. This function leverages the `__hash__` function of class `State` to store and retrieve states stored in the cache.

[23] Wikipedia Contributors. Python syntax and semantics. *Wikipedia, The Free Encyclopedia*. Available at: https://en.wikipedia.org/wiki/Python_syntax_and_semantics#Decorators

Listing 49 Memoization utility.

```
import functools

class memoize(object):
  """
  Memoization utility
  """

  def __init__(self, func):
    self.func, self.memoized, self.method_cache = func, {}, {}

  def __call__(self, *args):
    return self.cache_get(self.memoized, args, lambda: self.func(*args))

  def __get__(self, obj, objtype):
    return self.cache_get(self.method_cache, obj,
      lambda: self.__class__(functools.partial(self.func, obj)))

  def cache_get(self, cache, key, func):
    try:
      return cache[key]
    except KeyError:
      cache[key] = func()
      return cache[key]

  def reset(self):
    self.memoized, self.method_cache = {}, {}
```

A SAMPLE INSTANCE is presented in Listing 50. The solution to this instance is of course the same already illustrated in Fig. 41.

Listing 50 Capacitated lot sizing, sample instance.

```
instance = {"K": 40, "v": 1, "h": 1, "d":[10,20,30,40], "I0": 0, "C": 30}
CapacitatedLotSizingSDP(**instance)
```

WHEN THE CAPACITY IS THE SAME IN EVERY PERIOD, the problem has polynomial $O(T^4)$ complexity.[24]

[24] Michael Florian and Morton Klein. Deterministic production planning with concave costs and capacity constraints. *Management Science*, 18(1): 12–20, 1971.

Demand Forecasting

https://doi.org/10.11647/OBP.0252.03

Introduction

In this chapter, we discuss predictive analytics techniques for demand forecasting in inventory control. The techniques surveyed in this chapter originate in the realm of time series analysis and forecasting. We first introduce the notion of time series, then we survey a portfolio of time series models. We show how to fit these models to data and how to generate forecasts, confidence, and prediction bands.

Topics

Time series

Definition 4. *A time series is a series*

$$\{x_1, x_2, \dots\}$$

of indexed data points.

Examples of time series include: hourly temperatures at a given day/location, daily closing values of the Dow Jones Industrial Average, quarterly gas meter readings for a given household, etc.

Example 16. *In Table 4 we present a time series: the value at market close of the Dow Jones index between Mon 3 Aug 2020 and Fri 14 Aug 2020.*[25]

[25] https://finance.yahoo.com/

Date	Dow Jones
Mon 3 Aug 2020	26664.4
Tue 4 Aug 2020	26828.5
Wed 5 Aug 2020	27201.5
Thu 6 Aug 2020	27387.0
Fri 7 Aug 2020	27433.5
Mon 10 Aug 2020	27791.4
Tue 11 Aug 2020	27686.9
Wed 12 Aug 2020	27976.8
Thu 13 Aug 2020	27896.7
Fri 14 Aug 2020	27931.0

Table 4 Dow Jones Industrial Average between Mon 3 Aug 2020 and Fri 14 Aug 2020.

Time series analysis aims at extracting statistics and/or other information from time series data. The process typically starts with a so-called **exploratory analysis**, which aims at summarising key characteristics of time series data, often with visual methods, in order to formulate and test hypotheses.

Example 17. *In Fig. 42 we illustrate the behaviour of the Dow Jones Industrial Average between Mon 3 Aug 2020 and Fri 14 Aug 2020.*

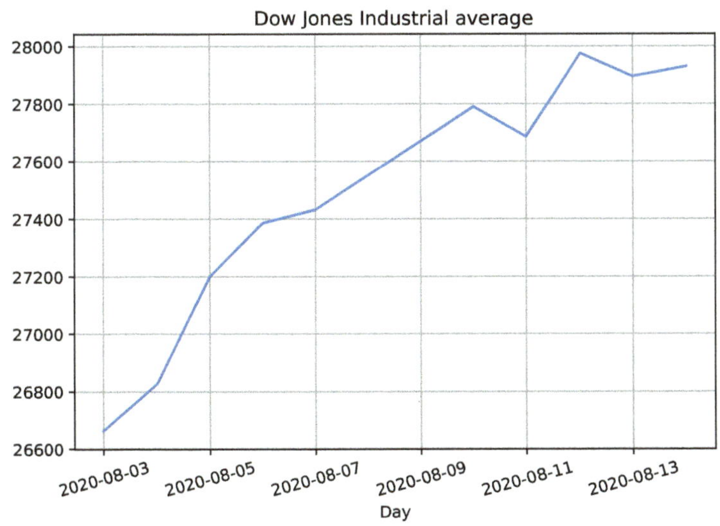

Fig. 42 Line chart of the Dow Jones Industrial Average between Mon 3 Aug 2020 and Fri 14 Aug 2020; note that there are no readings during the weekend, since the stock market is closed.

Time series forecasting leverages a *model*, e.g. a stochastic model, to predict future values based on previously observed values.

A possible approach to time series analysis and forecasting is to assume the time series is *a realisation* — i.e. an indexed set of observed values — of a given **stochastic process**.[26]

[26] Robert G. Gallager. *Stochastic processes*. Cambridge Univ. Pr., 2013.

Definition 5. *A stochastic process $\{X_t\}$ is an indexed set of random variables, where $t \in \mathcal{T}$, and the set $\mathcal{T}$ used to index the random variables is called the index set.*

A stationary stochastic process is a stochastic process whose unconditional joint probability distribution does not change when shifted in time.

Definition 6. *Let $\{X_t\}$ be a stochastic process; and $F_{t,\dots,n}(x)$ be the joint cumulative distribution function of $\{X_t, X_{t+1}, \dots, X_{t+n}\}$, where $n > 0$. $\{X_t\}$ is stationary if $F_{t,\dots,n}(x) = F_{t+\tau,\dots,n+\tau}(x)$, for all τ.*

Lemma 26. *A stochastic process $\{X_t\}$ in which all X_t are independent and identically distributed random variables is stationary.*

Definition 7 (White noise). *A stochastic process is said to be a white noise if its constituting random variables each have a probability distribution with zero mean and finite variance, and are mutually independent.*

Lemma 27. *A white noise is stationary.*

Definition 8 (Gaussian noise). *A Gaussian noise is a white noise in which all components follow a normal distribution with zero mean and the same variance σ^2.*

Example 18 (Gaussian white noise). *In Fig. 43 we illustrate 30 realisations of a standard Gaussian noise, i.e. $\sigma^2 = 1$.*

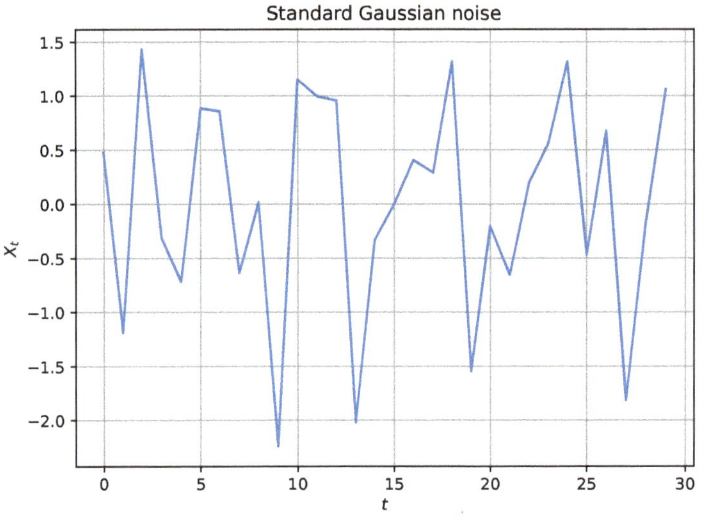

Fig. 43 A standard Gaussian noise.

Time series analysis techniques that assume the existence of an underpinning stochastic process may be divided into parametric and non-parametric.

Parametric approaches assume that there exists an underlying stationary stochastic process *possessing a certain structure*, which can be described using a small number of parameters; the task is then to estimate the parameters of the model that describes the stochastic process.

Non-parametric approaches do not assume that the underpinning stochastic process has any particular structure.

In what follows, we will focus on parametric approaches. Once a model for the underlying stochastic process has been chosen, one can carry out forecasting and predict the future behaviour of the underlying stochastic process.

Four simple forecasting methods

Behind a forecasting method there is often an underpinning working hypothesis that justifies it, and that motivates an underlying stochastic model. In this section, we will focus on four simple forecasting methods and on their underlying stochastic models.[27]

[27] The four methods are: the Moving Average method; the Naïve method; the Drift method; and the Seasonal Naïve method. It is often the case that one of these simple methods may be the best forecasting method available for a given application; alternatively, these methods may be used as benchmarks.

Stationary demand & the Moving Average method

The working hypothesis here is that: "tomorrow will be *roughly* the same as today." This hypothesis leads to a (stationary) stochastic process $\{X_t\}$ in which all X_t are independent and identically distributed random variables. Moreover, it is customary to assume forecast errors are normally distributed, therefore we will consider a Gaussian process as the underpinning stochastic process. To characterise this stochastic process, we must therefore know, or estimate, the distribution of X_t, that is its mean μ and standard deviation σ. In Listing 51 we show how to sample a Gaussian process with given mean μ and standard deviation σ in Python.

```
def sample_gaussian_process(mu, sigma,
      realisations):
   np.random.seed(1234)
   return np.random.normal(mu, sigma,
        realisations)
```

Listing 51 Sampling a Gaussian process in Python.

MOVING AVERAGE. This method predicts that

$$\widehat{X}_{t+k} \triangleq (x_{t-w+1} + \ldots + x_t)/t$$

for all $k = 1, 2, \ldots$; where $\widehat{X}_{t+k} \approx \mu$. In essence, the mean μ of all future random variables is assumed to be equal to the average of all historical realisations (Average method), or to the average of the past w realisations (Moving Average method). The following imports will be used throughout this section.

```
import math, statistics, scipy.stats as stats, statsmodels.api as sm
import numpy as np, pandas as pd
import matplotlib.pyplot as plt, pylab as py
```

We implement the Moving Average method in Python as follows.

```
def moving_average(series, w, t):
   """Forecasts elements t+1, t+2, ... of series
   """
   forecasts = np.empty(t+1)
   forecasts.fill(np.nan)
   for k in range(1,len(series)-t):
      forecasts = np.append(forecasts, series[t+1-w:t+1].mean())
   return forecasts
```

```
def plot(realisations, forecasts,
     window):
   f = plt.figure(1)
   plt.title("Moving Average
        forecasts\n window size =
        {}".format(window))
   plt.xlabel('Period ($t$)')
   first, last = next(x for x, val in
        enumerate(forecasts) if
        ~np.isnan(val)),
        len(forecasts)-1
   plt.axvspan(first, last, alpha=0.2,
        color='blue')
   plt.plot(forecasts, "g",
        label="Moving Average
        forecasts ($\widehat{X}_t$)")
   plt.plot(realisations,
        label="Actual values ($x_t$)")
   plt.legend(loc="upper left")
   plt.grid(True)
   f.show()
```

Listing 52 Plotting Moving Average forecasts in Python.

Example 19. *Consider a stochastic process $\{X_t\}$ where, for all t, random variable X_t is normally distributed with mean $\mu = 20$ and standard deviation $\sigma = 5$; we sample 200 realisations from $\{X_t\}$ and compute forecasts for the last 40 periods by using the Moving Average method with a window of size $w = 32$ (Fig. 44); Listing 52 illustrates the* `plot` *function.*

```
N, t, window = 200, 160, 32
realisations = pd.Series(sample_gaussian_process(20, 5, N), range(N))
forecasts = moving_average(realisations, window, t)
plot(realisations, forecasts, window)
py.show()
```

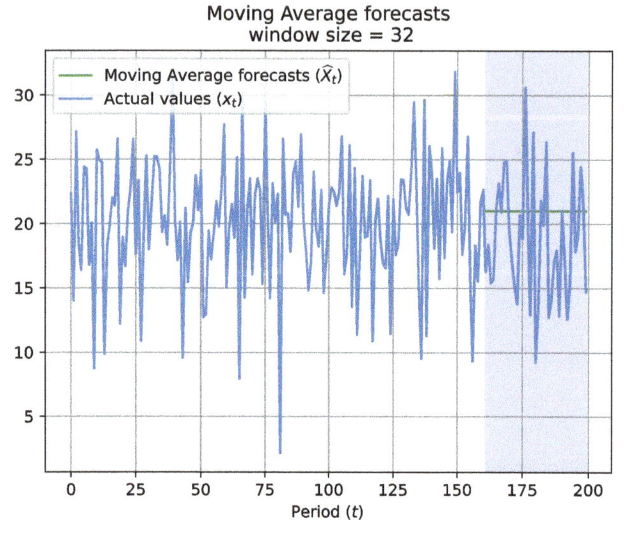

Fig. 44 Forecasts for the last 40 periods by using the Moving Average method with a window of size $w = 32$; the underpinning stochastic process is a a Gaussian process with mean $\mu = 20$ and standard deviation $\sigma = 5$.

Residuals Analysis. Recall that $x_1, x_2, \ldots, x_T$ are the realisations of $X_1, X_2, \ldots, X_T$. Consider realisations $x_1, x_2, \ldots, x_t$, and the one-step forecast $\widehat{X}_{t+1}$ given all previous realisations $x_1, x_2, \ldots, x_t$. For the Moving Average method, the list of all one-step forecasts for periods $1, \ldots, T$ is computed as follows.

```
def moving_average_rolling(series, w):
    return series.rolling(window=w).mean()
```

The residual $e_{t+1} = x_{t+1} - \widehat{X}_{t+1}$ represents the difference between realisation x_{t+1} and its one-step forecast $\widehat{X}_{t+1}$ based on all previous realisations $x_1, x_2, \ldots, x_t$. Residuals are what is left over after fitting a time series model. Given the list of all one-step forecasts, residuals can be computed as follows.

```
def residuals(realisations, forecasts):
    return realisations - forecasts
```

Residuals reveal if a model has adequately captured the information in the data. Let $e = \{e_1, \ldots, e_T\}$ be the residuals; a standardized residual can be computed by dividing the residual by the sample standard deviation of population e.

```
def standardised_residuals(realisations, forecasts):
    residuals = realisations - forecasts
    return (residuals) / statistics.stdev(residuals)
```

A good forecasting method will yield standardized residuals that have zero mean and are uncorrelated; ideally, residuals must approximate as closely as possible a standard Gaussian noise with constant variance (homoskedastic) over time periods. The following functions can be used for an exploratory analysis of residuals.

```
def residuals_plot(residuals):
    f = plt.figure(2)
    plt.xlabel('Period ($t$)')
    plt.ylabel('Residual')
    plt.plot(residuals, "g", label="Residuals")
    plt.grid(True)
    f.show()
```

```
def residuals_histogram(residuals):
    f = plt.figure(3)
    plt.xlabel('Residual')
    plt.ylabel('Frequency')
    num_bins = 30
    plt.hist(residuals, num_bins, facecolor='blue', alpha=0.5, density=True)
    x = np.linspace(-3, 3, 100)
    plt.plot(x, stats.norm.pdf(x, 0, 1))
    f.show()

def residuals_autocorrelation(residuals, window):
    f = plt.figure(4)
    plt.xlabel('Time lag')
    plt.ylabel('Autocorrelation')
    plt.acorr(residuals, maxlags=window) # autocorrelation of the residuals
    f.show()
```

The following code can be used to carry out a residuals analysis in Python for the Moving Average method.

```
N, window = 200, 32
realisations = pd.Series(sample_gaussian_process(20, 5, N), range(N))
forecasts = moving_average_rolling(realisations, window)
residuals = residuals(realisations[window:], forecasts[window:])
print("E[e_t] = "+str(statistics.mean(residuals)))
standardised_residuals = standardised_residuals(realisations[window:],
      forecasts[window:])
residuals_plot(residuals)
residuals_histogram(standardised_residuals)
residuals_autocorrelation(residuals, None)
sm.qqplot(standardised_residuals, line ='45')
py.show()
```

Residuals have mean -0.14, which is close to zero (Fig. 45).

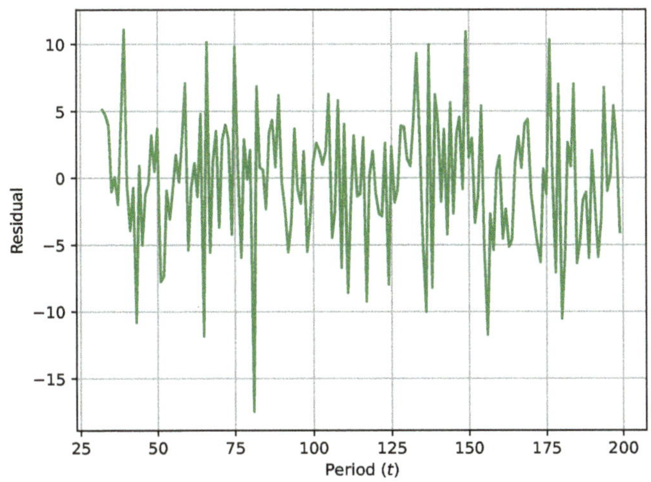

Fig. 45 Residual analysis for the Moving Average method: residuals.

The histogram in Fig. 46 suggests that residuals are approximately Gaussian. Fig. 47 reveals absence of residuals autocorrelation. Finally, the Q-Q plot (Fig. 48) appears to further support normality of residuals. These results suggests that the model has adequately captured the information in the data.

Besides analysing these property visually, one would generally also carry out statistical tests to test significance of these hypothesis. We direct the reader to the broader literature for more details.[28]

[28] Rob J. Hyndman and George Athanasopoulos. *Forecasting: Principles and practice*. OTexts, 2020.

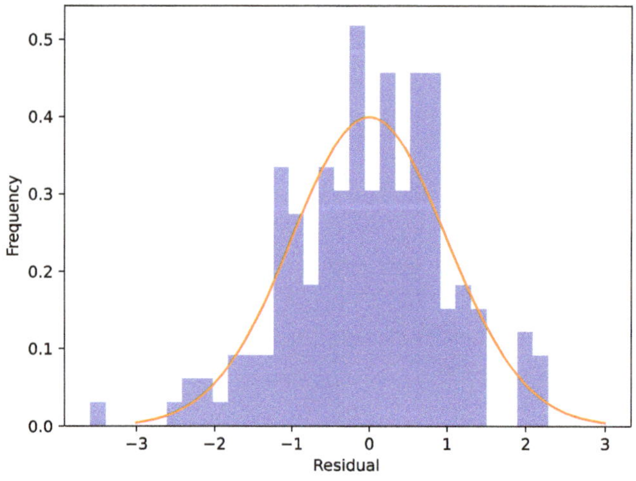

Fig. 46 Residual analysis for the Moving Average method: histogram.

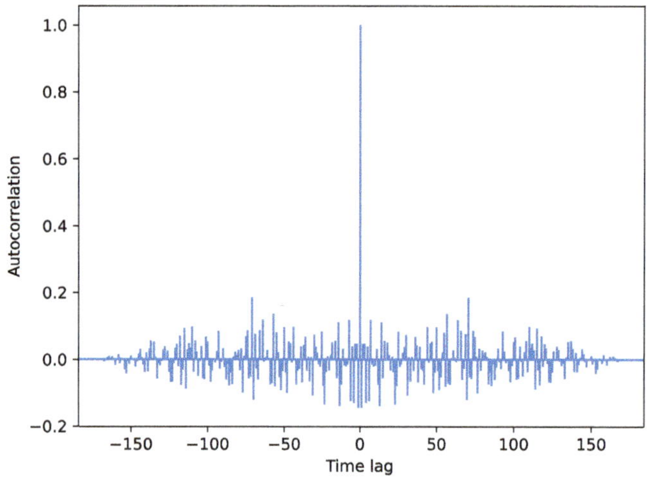

Fig. 47 Residual analysis for the Moving Average method: autocorrelation plot.

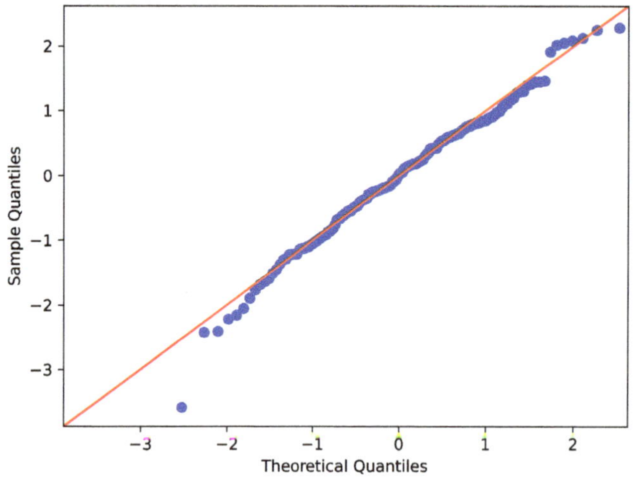

Fig. 48 Residual analysis for the Moving Average method: Q-Q plot.

Naïve method

The stochastic process of interest here is a *random walk*.

Definition 9. *A random walk is a stochastic process $\{X_t\}$ in which*

$$X_t = X_{t-1} + \varepsilon_t,$$

where stochastic process $\{\varepsilon_t\}$ is a white noise.

Random walk models are widely used for non-stationary data, particularly financial and economic data. They are often used when the change (i.e. difference) between consecutive observations in a given series appears to be a white noise: $x_t - x_{t-1} = \varepsilon_t$. Key characteristics of random walks are long periods of apparent trends up or down, paired with sudden and unpredictable changes in direction. Since future movements are unpredictable, and are equally likely to be up or down, the best forecast available is the last observation; this motivates the following forecasting method. In Listing 53 we show how to sample a random walk in Python.

```
def sample_random_walk(X0,
     realisations):
   np.random.seed(1234)
   errors = np.random.normal(0, 1,
       realisations)
   Xt = X0
   for e in errors:
      Xt = Xt + e
      yield Xt
```

Listing 53 Sampling a random walk in Python.

NAÏVE METHOD. The Naïve method, predicts that

$$\widehat{X}_{t+k} \triangleq x_t.$$

for all $k = 1, 2, \ldots$; in essence, the expected value of all future random variables is assumed to be equal to the value of the last observation. We next implement the Naïve method in Python.

```
def naive(series, t):
    """Forecasts periods t+1, t+2, ... of series
    """
    forecasts = np.empty(len(series))
    forecasts[:t+1] = np.nan
    forecasts[t+1:] = series[t]
    return forecasts
```

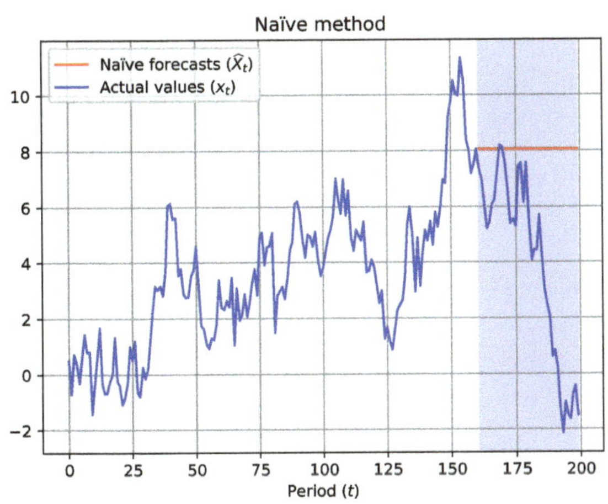

Fig. 49 Naïve method forecasts for the last 40 periods of a random walk with standard Gaussian noise.

Example 20. *Let $\{X_t\}$ be a random walk with standard Gaussian noise $\{\varepsilon_t\}$; by leveraging the following code, we sample 200 realisations from this process and compute the Naïve forecasts for the last 40 periods (Fig. 49). Listing 54 illustrates the* `plot` *function.*

```
N, t, window = 200, 160, 1
realisations = pd.Series(list(sample_random_walk(0, N)), range(N))
forecasts = naive(realisations, t)
plot(realisations, forecasts)
py.show()
```

```
def plot(realisations, forecasts):
    f = plt.figure(1)
    plt.title("Naive method")
    plt.xlabel('Period ($t$)')
    first, last = next(x for x, val in
        enumerate(forecasts) if
        ~np.isnan(val)),
        len(forecasts)-1
    plt.axvspan(first, last, alpha=0.2,
        color='blue')
    plt.plot(forecasts, "r",
        label="Naive forecasts
        ($\widehat{X}_t$)")
    plt.plot(realisations, "b",
        label="Actual values ($x_t$)")
    plt.legend(loc="upper left")
    plt.grid(True)
    f.show()
```

Listing 54 Plotting Naïve forecasts in Python for a random walk with standard Gaussian noise.

Naïve method one-step forecasts can be computed as follows.

```
def naive_rolling(series):
    return series.shift(periods=1)
```

By leveraging these forecasts, we carry out residuals analysis.

```
N, window = 200, 1
realisations = pd.Series(list(sample_random_walk(0, N)), range(N))
forecasts = naive_rolling(realisations)
residuals = residuals(realisations[window:], forecasts[window:])
print("E[e_t] = "+str(statistics.mean(residuals)))
standardised_residuals = standardised_residuals(realisations[window:],
    forecasts[window:])
residuals_plot(residuals)
residuals_histogram(standardised_residuals)
residuals_autocorrelation(residuals, None)
sm.qqplot(standardised_residuals, line ='45')
py.show()
```

Residuals have mean -0.009, which is close to zero (Fig. 50).

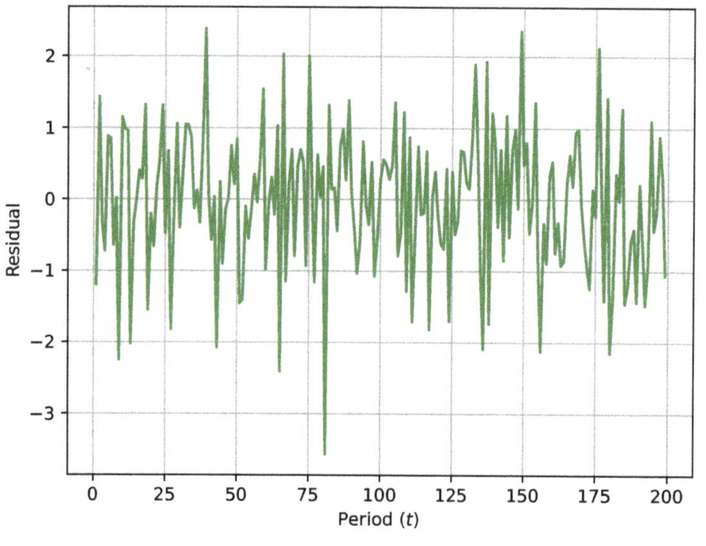

Fig. 50 Residual analysis for the Naïve method: residuals.

The histogram in Fig. 51 suggests that residuals are approximately Gaussian. Fig. 52 reveals absence of residuals autocorrelation. Finally, the Q-Q plot (Fig. 53) appears to further support normality of residuals.

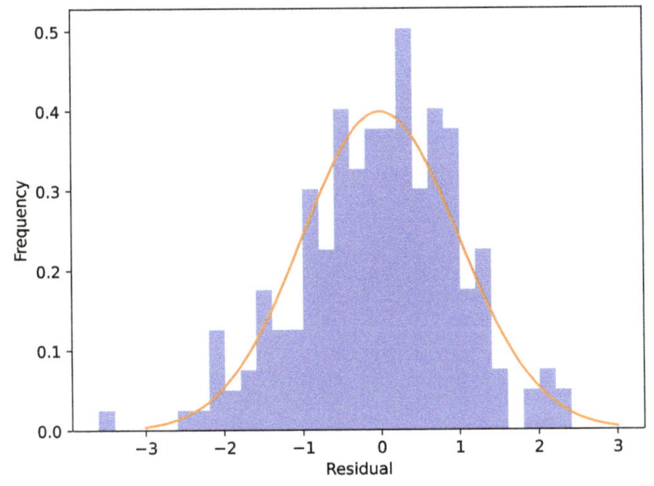

Fig. 51 Residual analysis for the Naïve method: histogram.

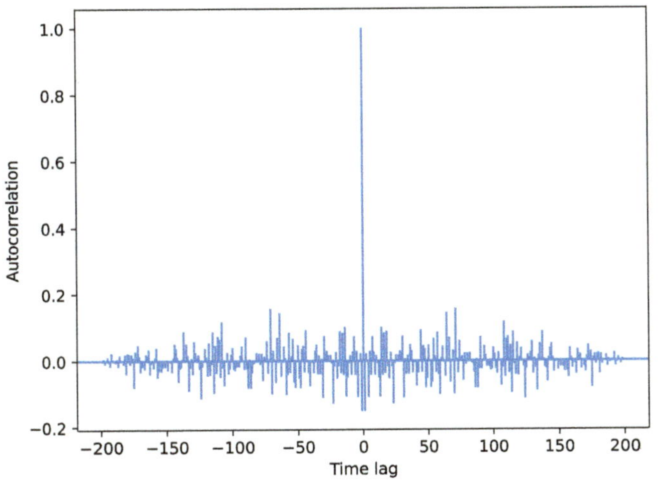

Fig. 52 Residual analysis for the Naïve method: autocorrelation plot.

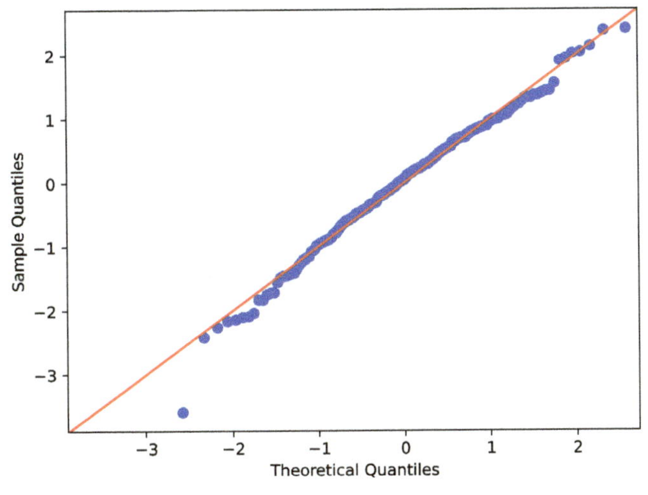

Fig. 53 Residual analysis for the Naïve method: Q-Q plot.

Drift method

A model closely related to the random walk allows the differences between consecutive observations to have a non-zero mean.

Definition 10. *A random walk with drift is a stochastic process* $\{X_t\}$ *such that*

$$X_t = c + X_{t-1} + \varepsilon_t,$$

where stochastic process $\{\varepsilon_t\}$ *is a white noise, and c is the drift.*

The "drift" represents the average change between consecutive observations, that is

$$\mathrm{E}[x_t - x_{t-1}] \approx c + \mathrm{E}[\varepsilon_t],$$

where $\mathrm{E}[\varepsilon_t] = 0$, since $\{\varepsilon_t\}$ is a white noise. If c is positive, $\{X_t\}$ will drift upwards; if c is negative, $\{X_t\}$ will drift downwards. In Listing 55 we sample a random walk with drift in Python.

```
def sample_random_walk(X0, c,
        realisations):
    np.random.seed(1234)
    errors = np.random.normal(0, 1,
        realisations)
    Xt = X0
    for e in errors:
        Xt = c + Xt + e
        yield Xt
```

Listing 55 Sampling a random walk with drift in Python.

DRIFT METHOD. The Drift method, predicts that

$$\widehat{X}_{t+k} \triangleq x_t + k/(t-1)\sum_{i=2}^{t}(x_i - x_{i-1}) = x_t + k(x_t - x_1)/(t-1)$$

for all $k = 1,2,\ldots$; this variant of the Naïve method allows the forecasts to increase or decrease over time, where the amount of change over time (the drift) is set to be the average change seen in the historical data. In essence, this is equivalent to drawing a line between the first and last observations, and extrapolating it into the future. The Drift method can be implemented in Python as follows.

```
def drift(series, t):
    """Forecasts periods t+1, t+2, ... of series
    """
    forecasts = np.empty(t+1)
    forecasts.fill(np.nan)
    x1 = series[0]
    xt = series[t]
    for k in range(1,len(series)-t):
        xtk = xt+k*(xt-x1)/t
        forecasts = np.append(forecasts, xtk)
    return forecasts
```

Example 21. *Let* $\{X_t\}$ *be a random walk with drift* $c = 0.1$ *and standard Gaussian noise* $\{\varepsilon_t\}$*; by leveraging the following code, we sample 200 realisations from this process and compute Drift forecasts for the last 40 periods (Fig. 54). Listing 56 illustrates the* `plot` *function.*

```
N, t, window = 200, 160, 2
realisations = pd.Series(list(sample_random_walk(0, 0.1, N)), range(N))
forecasts = drift(realisations, t)
plot(realisations, forecasts)
```

```
def plot(realisations, forecasts):
    f = plt.figure(1)
    plt.title("Drift method")
    plt.xlabel('Period ($t$)')
    first, last = next(x for x, val in
        enumerate(forecasts) if
        ~np.isnan(val)),
        len(forecasts)-1
    plt.axvspan(first, last, alpha=0.2,
        color='blue')
    plt.plot(forecasts, "r",
        label="Drift forecasts
        ($\widehat{X}_t$)")
    plt.plot(realisations, "b",
        label="Actual values ($x_t$)")
    plt.legend(loc="upper left")
    plt.grid(True)
    f.show()
```

Listing 56 Plotting Drift forecasts in Python for a random walk with standard Gaussian noise and drift.

Drift method one-step forecasts can be computed as follows.

```
def drift_rolling(series):
    forecasts = np.empty(2)
    forecasts.fill(np.nan)
    for k in range(2,len(series)):
        xk = drift(series[:k+1], k-1)[-1]
        forecasts = np.append(forecasts, xk)
    return forecasts
```

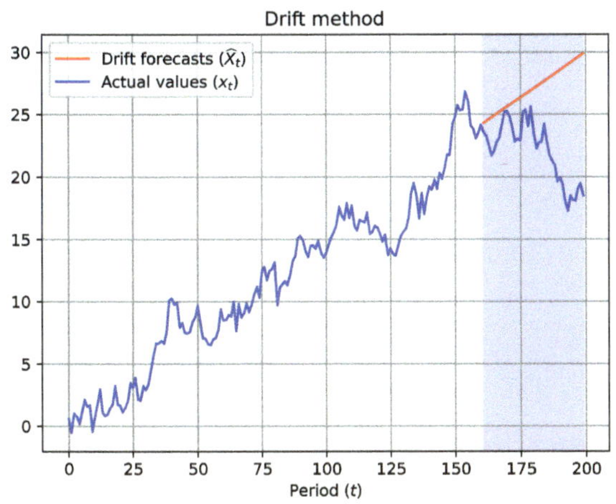

Fig. 54 Drift forecasts in Python for a random walk with standard Gaussian noise and drift $c = 0.1$.

By leveraging these forecasts, we carry out residuals analysis.

```
N, window = 200, 2
realisations = pd.Series(list(sample_random_walk(0, 0.1, N)), range(N))
forecasts = pd.Series(list(drift_rolling(realisations)), range(N))
residuals = residuals(realisations[window:], forecasts[window:])
print("E[e_t] = "+str(statistics.mean(residuals)))
standardised_residuals = standardised_residuals(realisations[window:],
      forecasts[window:])
residuals_plot(residuals)
residuals_histogram(standardised_residuals)
residuals_autocorrelation(residuals, None)
sm.qqplot(standardised_residuals, line ='45')
py.show()
```

Residuals have mean -0.02, which is close to zero (Fig. 55).

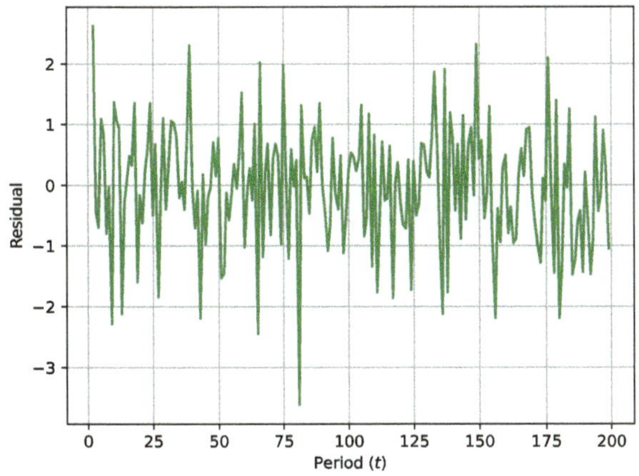

Fig. 55 Residual analysis for the Drift method: residuals.

The histogram in Fig. 56 suggests that residuals are approximately Gaussian. Fig. 57 reveals absence of residuals autocorrelation. Finally, the Q-Q plot (Fig. 58) appears to further support normality of residuals.

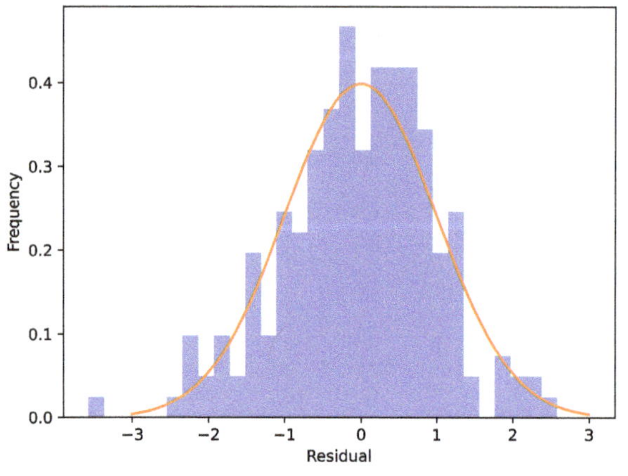

Fig. 56 Residual analysis for the Drift method: histogram.

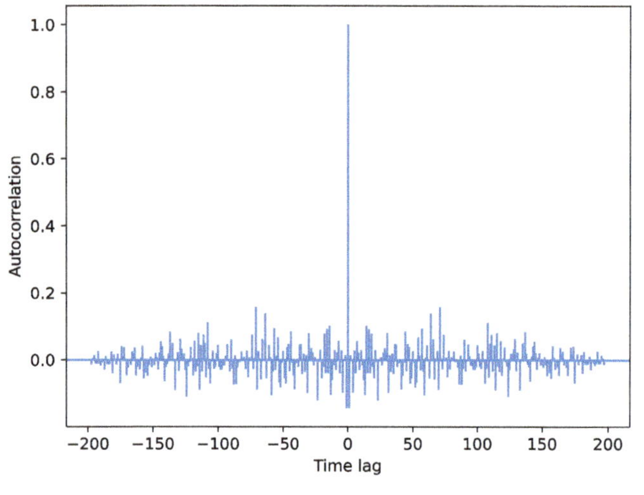

Fig. 57 Residual analysis for the Drift method: autocorrelation plot.

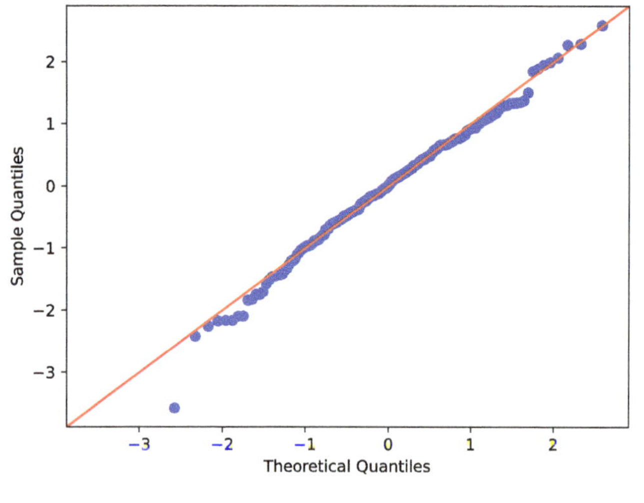

Fig. 58 Residual analysis for the Drift method: Q-Q plot.

Seasonal Naïve method

A seasonal difference is the difference between an observation and the previous observation from the same season, e.g. sales in November 2019 and sales in November 2020. A model closely related to the random walk considers the case in which seasonal differences in a given series appear to be a white noise. The stochastic process of interest is then a *seasonal random walk.*

Definition 11. *A seasonal random walk is a stochastic process $\{X_t\}$ in which*

$$X_t = X_{t-m} + \varepsilon_t,$$

where m is the number of seasons, and $\{\varepsilon_t\}$ is a white noise.

The differences $x_t - x_{t-m} = \varepsilon_t$ are called "lag-m differences." Since stochastic process $\{\varepsilon_t\}$ is a white noise, it follows that the average lag-m difference is assumed to be zero, that is

$$\mathrm{E}[x_t - x_{t-m}] \approx \mathrm{E}[\varepsilon_t] = 0.$$

In Listing 57 we show how to sample a seasonal random walk in Python.

```
def sample_seasonal_random_walk(
      realisations, m):
   np.random.seed(1234)
   errors = np.random.normal(0, 1,
         realisations)
   Xt = errors[:m]
   for t in range(m,realisations):
      Xt = np.append(Xt, Xt[t-m] +
            errors[t])
   return Xt
```

Listing 57 Sampling a seasonal random walk in Python.

SEASONAL NAÏVE METHOD. The Seasonal Naïve method, predicts that

$$\widehat{X}_{t+k} \triangleq x_{t+k-m(\lfloor (k-1)/m \rfloor + 1)}$$

for all $k = 1, 2, \ldots$; where m is the interval in periods between two "seasons," and $\lfloor x \rfloor$ rounds x down to the closest integer; for instance, assuming monthly data, the forecast for all future February values is equal to the last observed February value. This variant of the Naïve method allows seasonalities to be taken into account. The Seasonal Naïve method can be implemented in Python as follows.

```
def seasonal_naive(series, m, t):
   """Forecasts periods t+1, t+2, ... of series
   """
   forecasts = np.empty(len(series))
   forecasts[:t+1] = np.nan
   for k in range(t+1,len(series)):
      forecasts[k] = series[k-m*((k-t-1)//m+1)]
   return forecasts
```

Example 22. *Consider a stochastic process $\{X_t\}$ that is a seasonal random walk with $m = 5$ seasons and standard Gaussian noise $\{\varepsilon_t\}$; by leveraging the following code, we sample 100 realisations from this process and compute the Seasonal Naïve forecasts for the last 20 periods. (Fig. 59). Listing 58 illustrates the* `plot` *function.*

```
N, t, m = 100, 80, 5
realisations = pd.Series(list(sample_seasonal_random_walk(N, m)), range(N))
forecasts = seasonal_naive(realisations, m, t)
plot(realisations, forecasts)
py.show()
```

```
def plot(realisations, forecasts):
   f = plt.figure(1)
   plt.title("Seasonal naive method")
   plt.xlabel('Period ($t$)')
   first, last = next(x for x, val in
         enumerate(forecasts) if
         ~np.isnan(val)),
         len(forecasts)-1
   plt.axvspan(first, last, alpha=0.2,
         color='blue')
   plt.plot(forecasts, "r",
         label="Seasonal naive
         forecasts ($\widehat{X}_t$)")
   plt.plot(realisations, "b",
         label="Actual values ($x_t$)")
   plt.legend(loc="upper left")
   plt.grid(True)
   f.show()
```

Listing 58 Plotting Seasonal Naïve forecasts in Python for a seasonal random walk with standard Gaussian noise and $m = 5$ seasons.

Seasonal Naïve method one-step forecasts can be computed as follows.

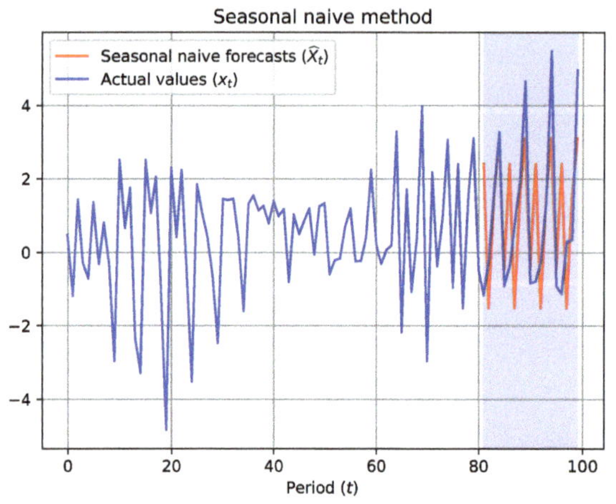

Fig. 59 Seasonal Naïve forecasts in Python for a seasonal random walk with standard Gaussian noise and $m = 5$ seasons.

```
def seasonal_naive_rolling(series, m):
   forecasts = np.empty(m)
   forecasts.fill(np.nan)
   for k in range(m,len(series)):
      xk = seasonal_naive(series[:k+1], m, k-1)[-1]
      forecasts = np.append(forecasts, xk)
   return forecasts
```

By leveraging these forecasts, we carry out residuals analysis.

```
N, m = 100, 5
realisations = pd.Series(list(sample_seasonal_random_walk(N, m)), range(N))
forecasts = pd.Series(list(seasonal_naive_rolling(realisations, m)), range(N))
residuals = residuals(realisations[m:], forecasts[m:])
print("E[e_t] = "+str(statistics.mean(residuals)))
standardised_residuals = standardised_residuals(realisations[m:], forecasts[m:])
residuals_plot(residuals)
residuals_histogram(standardised_residuals)
residuals_autocorrelation(residuals, None)
sm.qqplot(standardised_residuals, line ='45')
py.show()
```

Residuals have mean 0.04, which is close to zero (Fig. 60).

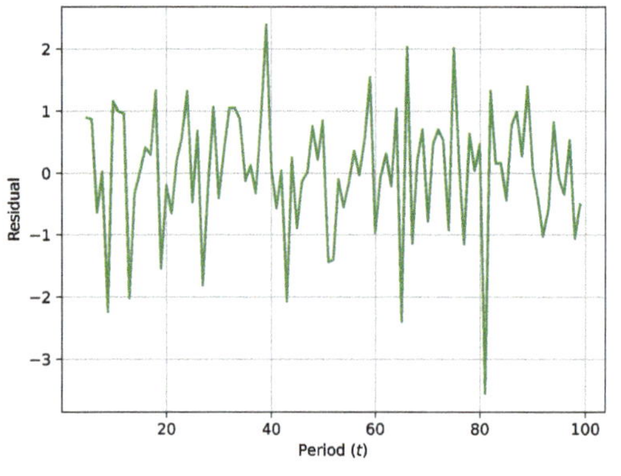

Fig. 60 Residual analysis for the Seasonal Naïve method: residuals.

The histogram in Fig. 61 suggests that residuals are approximately Gaussian. Fig. 62 reveals absence of residuals autocorrelation. Finally, the Q-Q plot (Fig. 63) appears to further support normality of residuals.

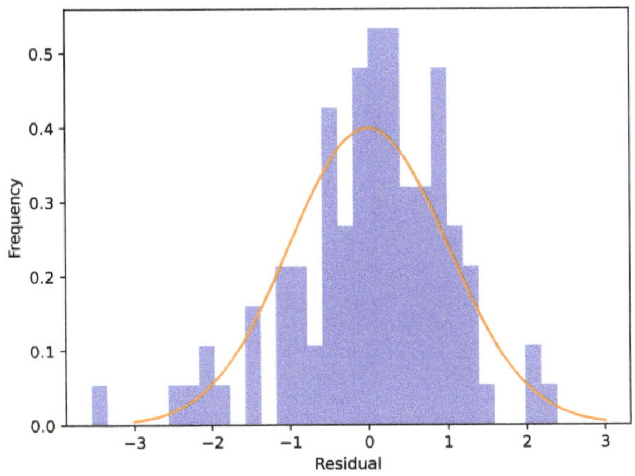

Fig. 61 Residual analysis for the Seasonal Naïve method: histogram.

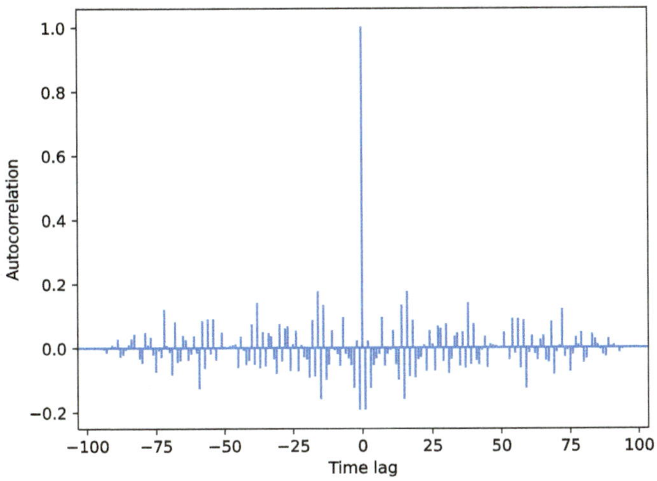

Fig. 62 Residual analysis for the Seasonal Naïve method: autocorrelation plot.

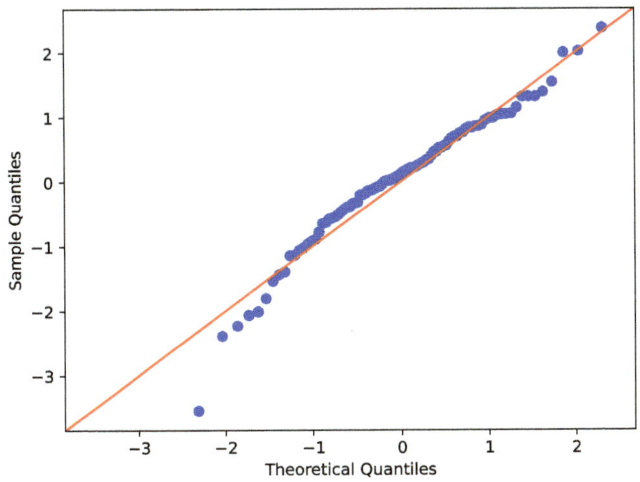

Fig. 63 Residual analysis for the Seasonal Naïve method: Q-Q plot.

Evaluating forecasting accuracy

Residuals are important to gauge the suitability of a fitted model, but they are not a reliable indication of how large true forecast errors are likely to be. The accuracy of forecasts can only be determined by considering how a model performs on new data that were not used when fitting the model.

Training vs testing

In forecasting, it is common practice to separate available data into training and test data. Training data are used to fit the forecasting model, while test data are used to evaluate the accuracy of the fitted model. Since test data are not used to fit the model, they can be used to assess how well the model may perform while forecasting new data.

A commonly adopted rule to separate training and test data is the 80/20 rule: 80% of the available sample will be devote to fitting the model, while the remaining 20% will be used to estimate forecast error (Fig. 65).

Fig. 64 Separating the available data into training and test data.

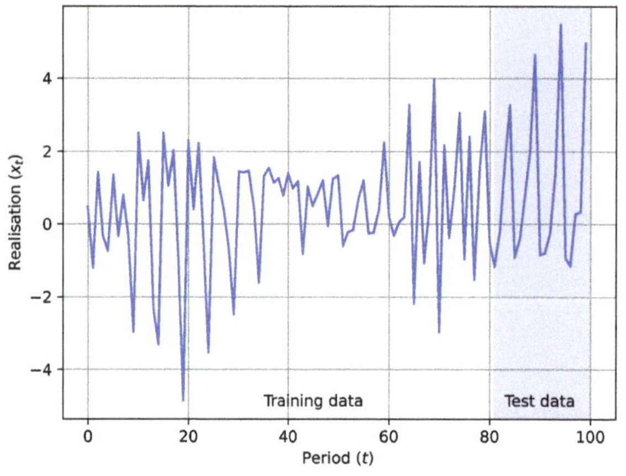

Forecast quality metrics

A forecast "error" is the difference between an observed value and its forecast. A forecast error does not denote a mistake; instead, it represents the random component of an observation.

Definition 12. *A forecast error is computed as*

$$\widehat{e}_{t+k} \triangleq x_{t+k} - \widehat{X}_{t+k},$$

where $\{x_1, \ldots, x_t\}$ *is the training set and* $\{x_{t+1}, x_{t+2} \ldots\}$ *is the test set.*

Forecast errors are different from residuals. First, residuals are calculated on the training set, while forecast errors are calculated on the test set. Second, residuals are based on one-step forecasts, while forecast errors can involve multi-step forecasts.

We can measure *forecasting accuracy* by summarising forecast errors in different ways as shown in Table 5.

Table 5 Forecast accuracy metrics.

Mean Absolute Error	$\mathrm{E}[\lvert e_t \rvert]$	`sklearn.metrics.mean_absolute_error`
Mean Squared Error	$\mathrm{E}[e_t^2]$	`sklearn.metrics.mean_squared_error`
Root Mean Squared Error	$\sqrt{\mathrm{E}[e_t^2]}$	`sqrt(mean_squared_error)`
Mean Absolute Percentage Error	$100\ \mathrm{E}[\lvert e_t \rvert / x_t]$	`mean_absolute_percentage_error`

$\mathrm{E}[x]$ denotes the expected value of x, and the function to compute the Mean Absolute Percentage Error is defined as follows.

```
def mean_absolute_percentage_error(y_true, y_pred):
    return np.mean(np.abs((y_true - y_pred) / y_true)) * 100}
```

Mean Absolute Error (MAE), Mean Squared Error (MSE), and Root Mean Squared Error (RMSE) are **scale-dependent error measures.** These measures cannot be used to make comparisons between series that involve different units, since forecast errors are on the same scale as the data. MAE is popular as it is easy to both understand and compute. A forecast method that minimises the MAE will lead to forecasts of the median, while minimising the RMSE will lead to forecasts of the mean. Consequently, the RMSE is also widely used, despite being more difficult to interpret.

Mean Absolute Percentage Error (MAPE) is a **percentage error measure**; it has the advantage of being unit-free, and is frequently used to compare forecast performances between data sets. Unfortunately, measures based on percentage errors have the disadvantage of being infinite or undefined if the realised value of the series is zero; and having extreme values if any realisation is close to zero. Moreover, it assumes the unit of measurement has a meaningful zero, and thus would not make sense if, say, we are measuring temperature in Fahrenheit or Celsius, because the position of the zero is arbitrary on these scales.

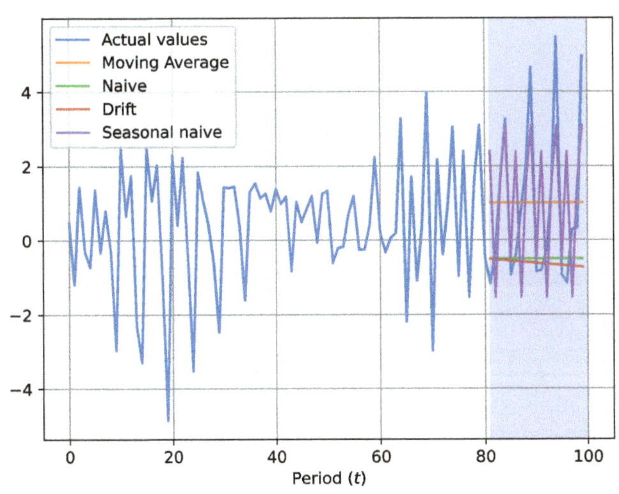

Fig. 65 All forecasting methods surveyed so far applied to a seasonal random walk with standard Gaussian noise and $m = 5$.

Example 23. *Consider a stochastic process $\{X_t\}$ that is a seasonal random walk with $m = 5$ seasons and standard Gaussian noise $\{\varepsilon_t\}$; we sample 100 realisations from this process and compute forecasts for the last 20 periods using all methods surveyed so far: Moving Average, Naïve, Drift, and Seasonal Naïve (Fig. 65). In Table 6 we report forecast accuracy metrics for all methods surveyed so far. These have been computed by using the following Python code.*

```
# training on 0..t
# testing on t+1,...N

N, t, window, m, test_window = 100, 80, 5, 5, [81,100]
realisations = pd.Series(list(sample_seasonal_random_walk(N, m)), range(N))
sma_forecasts = moving_average(realisations, window, t)
naive_forecasts = naive(realisations, t)
drift_forecasts = drift(realisations, t)
seasonal_naive_forecasts = seasonal_naive(realisations, m, t)

methods = {
  "Moving Average": sma_forecasts,
  "Naive": naive_forecasts,
  "Drift": drift_forecasts,
  "Seasonal naive": seasonal_naive_forecasts}

print("MAE")
for k in methods:
  print(k,end=':\t')
  print(mean_absolute_error(realisations[t+1:],methods[k][t+1:]))

print("\nMSE")
for k in methods:
  print(k,end=':\t')
  print(mean_squared_error(realisations[t+1:],methods[k][t+1:]))

print("\nRMSE")
for k in methods:
  print(k,end=':\t')
  print(math.sqrt(mean_squared_error(realisations[t+1:],methods[k][t+1:])))

print("\nMAPE")
for k in methods:
  print(k,end=':\t')
  print(mean_absolute_percentage_error(realisations[t+1:],methods[k][t+1:]))
```

The Seasonal Naïve forecasting method is known to be the optimal forecasting strategy for a seasonal random walk. MAE, MSE, and RMSE reflect this; in fact, they return the lowest scores for this method. However, MAPE scores are odd: not only they are large, but they seem to suggest that a Naïve method is the best performing forecasting strategy. This is due to the fact that several realisations for the underpinning stochastic process are close to zero; therefore, as previously mentioned, an MAPE will return extreme values, which in this instance are unreliable. We also know that there is no drift in the underpinning time series, and in fact the Naïve method outperforms the Drift method according to MAE, MSE, and RMSE. Performance of Moving Average and Naïve method is mixed, and there is no clear winner.

	MAE	MSE	RMSE	MAPE
Moving Average	1.77	4.47	2.11	190
Naïve	1.75	6.58	2.56	109
Drift	1.79	7.00	2.64	119
Seasonal Naïve	1.52	3.63	1.90	236

Table 6 Forecast accuracy metrics for different forecasting methods applied to a seasonal random walk with $m = 5$ seasons and standard Gaussian noise.

Prediction Intervals

Consider a time series $\{x_1, x_2, \ldots\}$, and assume that this time series has been generated by an underpinning stochastic process $\{X_t\}$. Recall that the so-called Average method operates under the assumption that random variables X_t are independent and identically distributed. In other words, it is assumed that the mean μ of all future random variables is equal to the mean of the random variables from which historical realisations have been drawn.

A first question one may want to address is to estimate the value of μ on the basis of past realisations. This question can be answered via *confidence interval analysis*.

Confidence intervals

Let $\{x_1, x_2, \ldots, x_n\}$ be a set of n independent realisations drawn from a random variable X with mean μ and variance σ^2, both of which are assumed to be unknown. Let

$$\bar{\mu} = (x_1 + x_2 + \ldots + x_n)/n$$

be the sample mean, and

$$\bar{\sigma}^2 = n(n-1)^{-1}\sqrt{(x_1^2 + x_2^2 + \ldots + x_n^2)/n - \bar{\mu}^2}$$

be the sample variance, where term $n(n-1)^{-1}$ is Bessel's correction,[29] which is necessary to obtain an unbiased estimate of the population variance from a finite sample of n observations.

[29] Douglas C. Montgomery and George C. Runger. *Applied statistics and probability for engineers*. John Wiley and Sons, 2014.

Definition 13. *The α confidence interval of the mean μ of X is*

$$\mathcal{I}(\alpha) \triangleq (\bar{\mu} - z\bar{\sigma}/\sqrt{n}, \bar{\mu} + z\bar{\sigma}/\sqrt{n})$$

where z is the $1-(1-\alpha)/2$ quantile of the inverse t distribution with $n-1$ degrees of freedom.

Lemma 28. *With confidence probability α, the α confidence interval of the mean will cover the mean μ.*

We next show a simple Python code to illustrate the concept of confidence interval coverage (Fig. 66).

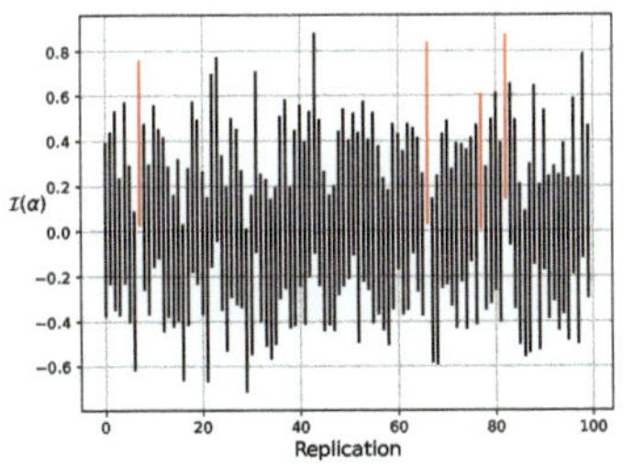

Fig. 66 Estimation of the mean μ. Confidence intervals ($\alpha = 0.95$) have been computed for 100 replications of $n = 30$ realisations drawn from a standard normal random variable. In 4 instances out of 100 (approx. 95%), the interval did not cover the true mean value $\mu = 0$; these instances are marked in red.

```
import math
import numpy as np
from scipy.stats import t
import statistics as s
import matplotlib.pyplot as plt

np.random.seed(1234)
replications = 100
n = 30
x = range(replications)
y = np.random.normal(0, 1, size=(replications, n)) # realisations
alpha = 0.95 # confidence level
z = t.ppf(1-(1-alpha)/2, n-1) # inverse t distribution
y_mean = [s.mean(y[r]) for r in range(replications)]
e = [z*s.stdev(y[r])/math.sqrt(n) for r in range(replications)]
ec = ['red' if (y_mean[r]+z*s.stdev(y[r])/math.sqrt(n) < 0 or
          y_mean[r]-z*s.stdev(y[r])/math.sqrt(n) > 0)
      else 'black' for r in range(replications)]
plt.errorbar(x, y_mean, yerr=e, ecolor=ec, fmt='none')
plt.grid(True)
plt.show()
```

Whilst it is interesting to estimate the value of μ for a stationary stochastic process, in forecasting what we would really like to know is, given a set of past observations, the interval within which we expect the next observation(s) to lie with a specified probability. This interval is called the *prediction interval*. In particular, we talk about *one-step* prediction intervals, if we are forecasting one step ahead; and of *multi-step* prediction intervals, if we are forecasting multiple periods ahead.

Prediction intervals

Let us consider a stochastic process $\{X_t\}$ such that, for all t, X_t are independent and identically distributed (iid) normal random variables with known mean μ and standard deviation σ.

Definition 14. *The α prediction interval of a future realisation x_{n+1} of random variable X_{n+1} is*

$$(\mu - z\sigma, \mu + z\sigma)$$

where z is the $1-(1-\alpha)/2$ quantile of an inverse standard normal distribution.

Lemma 29. *With probability α, a future realisation x_{n+1} of random variable X_{n+1} falls within this interval.*

We next show a simple Python code to illustrate the concept of prediction interval coverage (Fig. 67).

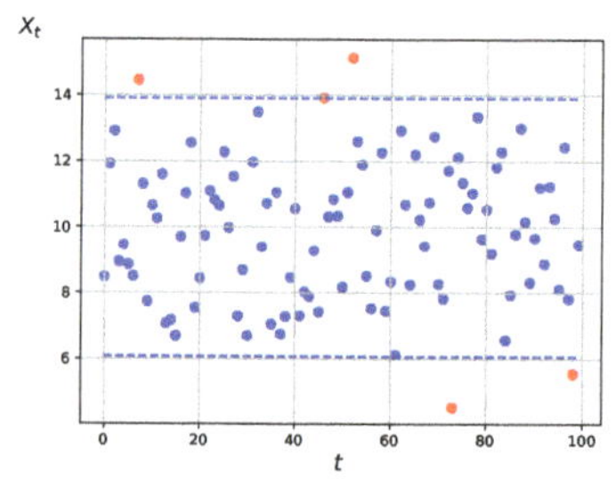

Fig. 67 Prediction intervals ($\alpha = 0.95$) of a gaussian stochastic process $\{X_t\}$, where, for all t, X_t is a normal random variable with $\mu = 10$ and $\sigma = 2$. In 5 instances out of 100 (95%), a realisation did not fall within the prediction interval; these instances are marked in red.

```
from matplotlib import colors
import numpy as np
from scipy.stats import norm
import matplotlib.pyplot as plt

np.random.seed(4321)
replications = 100
x = range(replications)
mu, sigma = 10, 2
y = np.random.normal(mu, sigma, replications) # realisations
alpha = 0.95 # confidence level
z = norm.ppf(1-(1-alpha)/2) # inverse standard normal distribution
plt.plot(x,[mu-z*sigma for k in x], color='blue', linestyle='dashed')
plt.plot(x,[mu+z*sigma for k in x], color='blue', linestyle='dashed')
ec = ['red' if y[r]>mu+z*sigma or y[r]<mu-z*sigma
         else 'blue' for r in range(replications)]
plt.scatter(x,y,color=ec)
plt.grid(True)
plt.show()
```

It is worth observing that the α prediction interval of a future realisation x_t of random variable X_t is independent of t.[30]

[30] Note that for some forecasting methods, this will not necessarily be the case.

Let us now assume that we are given $\{x_1, x_2, \ldots, x_n\}$ realisations, and that our aim is to compute the α prediction interval for a future observation X_{n+1} of stochastic process $\{X_t\}$. However, now μ and σ are unknown and must be estimated from past realisations. A possible heuristic may be to replace the unknown mean μ and standard deviation σ with the sample mean $\bar{\mu}$ and sample variance $\bar{\sigma}$, respectively; and then apply the approach just outlined. However, the resulting intervals will not be prediction intervals, since future realisations will not be guaranteed to fall in it according to the prescribed probability α. We shall next see how to obtain prediction intervals when mean μ and standard deviation σ are unknown.

We shall begin by considering the case in which the mean μ is unknown and the standard deviation σ is known and equal to 1.

Unknown μ, known $\sigma = 1$

Lemma 30. *Consider stochastic process $\{X_t\}$ with unknown μ and $\sigma = 1$; the α prediction interval of a future realisation x_{n+1} of random variable X_{n+1} given realisations $\{x_1, x_2, \ldots, x_n\}$*

$$(\bar{\mu} - z\sqrt{1 + 1/n}, \bar{\mu} + z\sqrt{1 + 1/n})$$

where z is the $1 - (1 - \alpha)/2$ quantile of an inverse standard normal distribution.

Proof. Observe that the sample mean $\bar{\mu}$ is normally distributed with mean μ and standard deviation $\sigma = \sqrt{1/n}$, while the future observation X_{n+1} is normally distributed with mean μ and standard deviation 1. Then $X_{n+1} - \bar{\mu}$ is normally distributed with mean 0 and standard deviation $\sigma = \sqrt{1 + 1/n}$. The prediction distribution for X_{n+1} is therefore a normal distribution with mean $\bar{\mu}$ and standard deviation $\sigma = \sqrt{1 + 1/n}$. □

Next, we consider the case in which the mean μ is known and equal to zero, and the standard deviation σ is unknown.

Known $\mu = 0$, unknown σ

Lemma 31. *Consider stochastic process $\{X_n\}$ with $\mu = 0$ and unknown σ; the α prediction interval of a future realisation x_{n+1} of random variable X_{n+1} given realisations $\{x_1, x_2, \ldots, x_n\}$*

$$(-zs, zs)$$

where z is the $1 - (1 - \alpha)/2$ quantile of the inverse t distribution with $n - 1$ degrees of freedom.

Proof. The sample variance s^2 of $\{x_1, x_2, \ldots, x_n\}$, scaled by factor $(n - 1)/\sigma^2$, follows a χ^2 distribution with $n - 1$ degrees of freedom, while the future observation X_{n+1} is normally distributed with mean $\mu = 0$ and variance σ^2. If we take the ratio X_{n+1}/s, the two terms σ cancel out, and what remains is known to follow a Student's t-distribution with $n - 1$ degrees of freedom. The prediction distribution for X_{n+1}/s is therefore a Student's t-distribution with $n - 1$ degrees of freedom. □

Finally, we consider the case in which both the mean μ and the standard deviation σ are unknown.

Unknown μ, unknown σ

Lemma 32. *Consider stochastic process $\{X_t\}$ with unknown μ and unknown σ; the α prediction interval of a future realisation x_{n+1} of random variable X_{n+1} given realisations $\{x_1, x_2, \ldots, x_n\}$*

$$(\bar{\mu} - zs\sqrt{1 + 1/n}, \bar{\mu} + zs\sqrt{1 + 1/n})$$

where z is the $1 - (1 - \alpha)/2$ quantile of the inverse t distribution with $n - 1$ degrees of freedom.

Proof. The result follows by combining the two previous results for unknown μ, known $\sigma = 1$; and known $\mu = 0$, unknown σ. This combination is possible because the sample mean and sample variance of the normal distribution are independent statistics. □

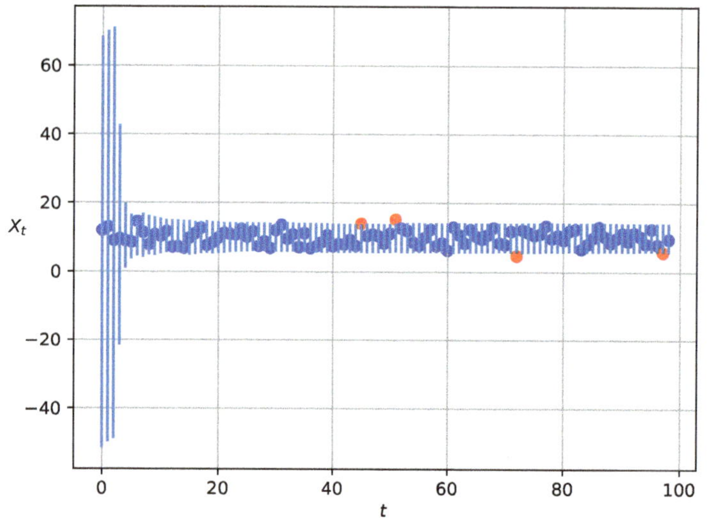

Fig. 68 Prediction intervals ($\alpha = 0.95$) of a gaussian stochastic process $\{X_t\}$, where, for all t, X_t is a normal random variable with unknown μ and σ. In 4 instances out of 100 (approx. 95%), a realisation did not fall within the prediction interval; these instances are marked in red.

We next show a simple Python code to illustrate prediction interval coverage for the case in which both the mean μ and the standard deviation σ are unknown (Fig. 68).

```
import math
import numpy as np
import statistics as s
from statistics import mean
from scipy.stats import t
import matplotlib.pyplot as plt
from matplotlib import colors

np.random.seed(4321)
replications, mu, sigma = 100, 10, 2
x = range(replications)
y = np.random.normal(mu, sigma, replications) # realisations
alpha = 0.95 # confidence level
z = lambda n: t.ppf(1-(1-alpha)/2, n-1) # inverse t distribution
y_mean = [s.mean(y[0:r+1]) for r in range(replications)]
e = [z(r-1)*s.stdev(y[0:r+1])*math.sqrt(1+1/r)
     if r > 2 else 30*sigma for r in range(replications)]
plt.errorbar(x[:-1], y_mean[:-1], yerr=e[:-1], fmt='none')
ec = ['red' if y[1:][r]>y_mean[:-1][r]+e[:-1][r] or
              y[1:][r]<y_mean[:-1][r]-e[:-1][r]
           else 'blue' for r in range(replications-1)]
plt.scatter(x[:-1], y[1:], color=ec)
plt.grid(True)
plt.show()
```

Observe that in Fig. 68 the size of the prediction intervals varies (in particular, it shrinks) with the number of past realisations that are available for the estimation.

We have shown how to compute prediction intervals for the case in which we are forecasting a stationary stochastic process $\{X_t\}$ where, for all t, X_t is a normal random variable with unknown μ and σ. This the stochastic process that underpins the Average Method. Therefore the prediction intervals presented apply to the Average method and the Moving Average method. Moreover, in this specific case, at period t the prediction interval of a future realisation x_{t+k} of random variable X_{t+k} is independent of k. Therefore one-step and multi-step prediction intervals coincide.

Similarly to what we have seen for the Average method, it is possible to derive prediction intervals for the other three benchmark methods previously presented: the Naïve method, the Seasonal Naïve method, and the Drift method. Consider realisations $\{x_1, x_2, \ldots, x_n\}$, let $\bar{\sigma}$ be the residuals standard deviation computed for a given method, and let $\bar{\sigma}_k$ denote standard deviation of the k-step forecast distribution. In Table 7 we summarise, for each method, the expressions of the n-step forecast distribution mean and standard deviation.

	$\widehat{X}_{n+k}$	$\bar{\sigma}_k$
Average	$(x_1 + \ldots + x_n)/n$	$\bar{\sigma}\sqrt{1+1/n}$
Naïve	x_n	$\bar{\sigma}\sqrt{k}$
Seasonal Naïve	$x_{t+k-m(\lfloor (k-1)/m \rfloor+1)}$	$\bar{\sigma}\sqrt{\lfloor (k-1)/m \rfloor + 1}$
Drift	$x_t + k(x_t - x_1)/(t-1)$	$\bar{\sigma}\sqrt{k(1+k/n)}$

Table 7 Expressions of the n-step forecast distribution mean and standard deviation.

The following code amends function `plot` previously presented for the Naïve method to display prediction intervals (Fig. 69).

```
def plot(realisations, forecasts, stdev, alpha):
    f = plt.figure(1)
    plt.title("Naive method")
    plt.xlabel('Period ($t$)')
    first, last = next(x for x, val in enumerate(forecasts) if ~np.isnan(val)),
           len(forecasts)-1
    plt.axvspan(first, last, alpha=0.2, color='blue')
    plt.plot(forecasts, "r", label="Naive forecasts ($\widehat{X}_t$)")
    plt.plot(realisations, "b", label="Actual values ($x_t$)")
    z = t.ppf(1-(1-alpha)/2, len(realisations)-1) # inverse t distribution
    plt.fill_between(range(first, last+1),
        [forecasts[first+k]-z*stdev*math.sqrt(k) for k in range(last-first+1)],
        [forecasts[first+k]+z*stdev*math.sqrt(k) for k in range(last-first+1)],
        color='r', alpha=0.1)
    plt.legend(loc="upper left")
    plt.grid(True)
    f.show()

N, t, window, alpha = 200, 160, 1, 0.95
realisations = pd.Series(list(sample_random_walk(0, N)), range(N))
forecasts = naive(realisations, t)
forecasts_roll = naive_rolling(realisations)
residuals = residuals(realisations[window:], forecasts_roll[window:])
plot(realisations, forecasts, s.stdev(res), alpha)
print("E[e_t] = "+str(s.mean(residuals)))
print("Stdev[e_t] = "+str(s.stdev(residuals)))
plt.show()
```

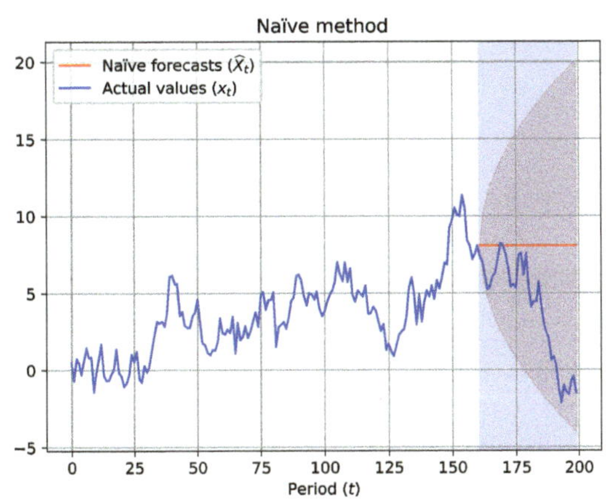

Fig. 69 Naïve method forecasts and prediction intervals for the last 40 periods of a random walk with standard Gaussian noise.

Box-Cox transformations

If the time series show variation that increases or decreases with the level of the series, then we can adopt logarithmic transformations or power transformations to stabilise variation.

A family of transformation that includes both logarithms and power transformations, is the family of Box-Cox transformations,[31] which depend upon a parameter λ. Consider a time series $y_1, y_2, \ldots$; a Box-Cox transformation is defined as

$$w_t = \begin{cases} \log(y_t) & \lambda = 0 \\ (y_t^\lambda - 1)/\lambda & \text{otherwise} \end{cases}$$

where w_t are the elements of the transformed series.

[31] George. E. P. Box and David R. Cox. An analysis of transformations. *Journal of the Royal Statistical Society: Series B (Methodological)*, 26(2):211–243, 1964.

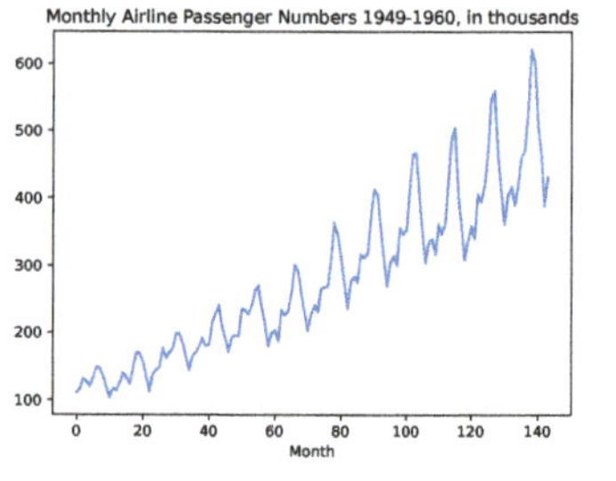

Fig. 70 Airline time series.

Example 24. *Consider the dataset of Monthly Airline Passenger Numbers*[32] *1949-1960, in thousands, obtained via the following Python code and shown in Fig. 70.*

[32] San Francisco Open Data portal, https://data.sfgov.org/.

```
import statsmodels.api as sm, pandas as pd
import matplotlib.pyplot as plt

airpass = sm.datasets.get_rdataset("AirPassengers", "datasets")
plt.title("Monthly Airline Passenger Numbers 1949-1960, in thousands")
plt.plot(pd.Series(airpass.data["value"]))
plt.show()
```

We now apply Box-Cox transformation as follows, and let the algorithm choose the best value of λ.

```
import scipy.stats as stats

airpass = sm.datasets.get_rdataset("AirPassengers", "datasets")
series, l = stats.boxcox(airpass.data["value"])
print("optimal lambda: "+str(l))
plt.plot(series)
plt.show()
```

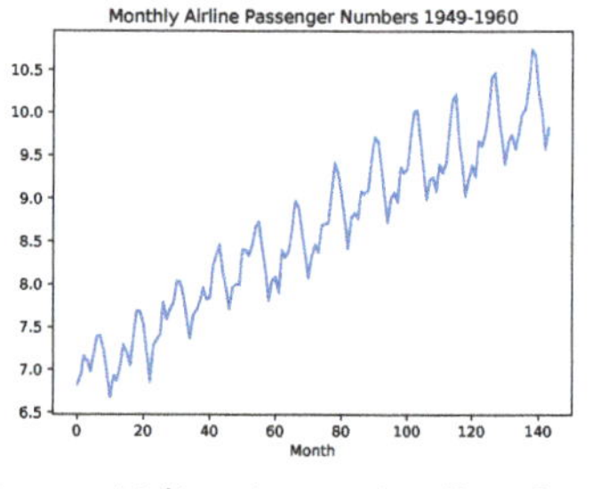

Fig. 71 Airline time series: Box-Cox transformation ($\lambda = 0.148$).

The transformed series is shown in Fig. 71; the optimal value of lambda chosen by the algorithm is $\lambda = 0.148$.

Having chosen a transformation, we forecast the transformed data. Then, we need to reverse the transformation (or back-transform) to obtain forecasts on the original scale. The reverse Box-Cox transformation is given by

$$y_t = \begin{cases} e^{y_t} & \lambda = 0 \\ (\lambda w_t + 1)^{1/\lambda} & \text{otherwise.} \end{cases}$$

In Python, this back-transformation is obtained via the following code

```
from scipy.special import inv_boxcox

series = inv_boxcox(series, l)
```

The inverted series is identical to the original series in Fig. 70.

Exponential Smoothing

Exponential smoothing was proposed in the late 1950s,[33] and has motivated some of the most successful forecasting methods.

[33] Robert G. Brown. *Statistical forecasting for inventory control*. McGraw-Hill, 1959.

Simple Exponential Smoothing

Consider a stochastic process $\{X_t\}$ and recall that, given realisations $\{x_1, x_2, \ldots, x_n\}$, the Average method predicts that

$$\widehat{X}_{n+k} \triangleq (x_1 + \ldots + x_n)/n$$

for all $k = 1, 2, \ldots$; where $\widehat{X}_{n+k} \approx \mu$. In essence, the mean μ of all future random variables is assumed to be equal to the average of all historical realisations.[34] We can rewrite this expression as

$$\widehat{X}_{n+k} = \alpha_1 x_1 + \alpha_2 x_2 + \ldots + \alpha_n x_n,$$

where $\alpha_1 = \alpha_2 = \ldots = \alpha_n = 1/n$. This evidences that all past realisations are given equal weight in the computation.

[34] We say that Simple Exponential Smoothing has a "flat" forecast function: all future forecasts take the same value. Of course, this means the method will only be suitable if the time series has no trend or seasonal component.

Instead of weighting equally all past realisations, in Simple Exponential Smoothing older realisations receive a weight that is exponentially smaller than that assigned to more recent realisations. Given realisations $\{x_1, x_2, \ldots, x_n\}$, the method predicts that

$$\widehat{X}_{n+k} \triangleq \alpha x_n + (1-\alpha)\widehat{X}_n \tag{28}$$

for all $k = 1, 2, \ldots$; where $\widehat{X}_n$ is the previous forecast[35] based on realisations $\{x_1, x_2, \ldots, x_{n-1}\}$; and $0 < \alpha < 1$ is the smoothing parameter.

[35] $\widehat{X}_n$ is also referred to as "the level" (or the smoothed value) of the series at time n.

By expanding Eq. 28 for $k = 1$ we obtain

$$\begin{aligned}
\widehat{X}_2 &= \alpha x_1 + (1-\alpha)x_0 \\
\widehat{X}_3 &= \alpha x_2 + (1-\alpha)\widehat{X}_2 \\
&\vdots \\
\widehat{X}_{n+1} &= \alpha x_n + (1-\alpha)\widehat{X}_n
\end{aligned}$$

where x_0 is an arbitrary constant denoting our initial estimate. Finally,

$$\begin{aligned}
\widehat{X}_2 &= \alpha x_1 + (1-\alpha)x_0 \\
\widehat{X}_3 &= \alpha x_2 + (1-\alpha)(\alpha x_1 + (1-\alpha)x_0) \\
&= \alpha x_2 + \alpha(1-\alpha)x_1 + (1-\alpha)^2 x_0 \\
\widehat{X}_4 &= \alpha x_3 + (1-\alpha)(\alpha x_2 + \alpha(1-\alpha)x_1 + (1-\alpha)^2 x_0) \\
&= \alpha x_3 + \alpha(1-\alpha)x_2 + \alpha(1-\alpha)^2 x_1 + (1-\alpha)^3 x_0 \\
&\vdots \\
\widehat{X}_{n+1} &= \sum_{j=0}^{n} \alpha(1-\alpha)^j x_j + (1-\alpha)^n x_0
\end{aligned}$$

Observe that term $(1-\alpha)^n x_0$ vanishes for large n; and also that $\alpha > \alpha(1-\alpha) > \alpha(1-\alpha)^2 > \ldots$; this means older realisations receive a weight that is exponentially smaller than that assigned to more recent realisations.

Based on the previous discussion, the stochastic process underpinning Simple Exponential Smoothing must be stationary, since the time series being forecasted must have no trend or seasonal component. This observation leads to the following hierarchy

Method	Description
Average	consider all past observations equally weighted
Moving Average	consider only the most recent w observations, equally weighted
Simple Exponential Smoothing	consider all past observations, older realisations receive a weight exponentially smaller than that assigned to more recent realisations

The key difference between the Average method, and its two variants (Moving Average & Simple Exponential Smoothing), is that these variants implement "forgetting" in two different forms, and thus try to discount past observations in one way or another. This may be appropriate for forecasting stochastic processes that are approximately stationary and/or change slowly over time.

We implement Simple Exponential Smoothing in Python as shown in Listing 59.

```
def ses(series, alpha, x0, t):
    """Forecasts elements t+1, t+2, ...
       of series
    """
    forecasts = np.empty(len(series))
    forecasts[0] = x0
    for k in range(1,t+2):
        forecasts[k] = alpha*series[k-1]
            + (1-alpha)*forecasts[k-1]
    for k in range(t+2,len(series)):
        forecasts[k] = forecasts[k-1]
    forecasts[0:t] = np.nan
    return forecasts
```

Listing 59 Simple Exponential Smoothing in Python.

Example 25. *Consider a stochastic process $\{X_t\}$ where, for all t, random variable X_t is normally distributed with mean $\mu = 20$ and standard deviation $\sigma = 5$; we sample 200 realisations from $\{X_t\}$ and compute forecasts for the last 40 periods by using Simple Exponential Smoothing with $\alpha = 0.5$ (Fig. 72); Listing 60 illustrates the `plot` function.*

```
N, t, alpha, x0 = 200, 160, 0.5, 20
realisations = pd.Series(sample_gaussian_process(20, 5, N), range(N))
forecasts = ses(realisations, alpha, x0, t)
plot(realisations, forecasts, alpha)
```

```
def plot(realisations, forecasts,
     alpha):
    f = plt.figure(1)
    plt.title("Simple Exponential
        Smoothing forecasts\n alpha =
        {}".format(alpha))
    plt.xlabel('Period ($t$)')
    first, last = next(x for x, val in
        enumerate(forecasts) if
        ~np.isnan(val)),
        len(forecasts)-1
    plt.axvspan(first, last, alpha=0.2,
        color='blue')
    plt.plot(forecasts, "g",
        label="Simple Exponential
        Smoothing forecasts
        ($\widehat{X}_t$)")
    plt.plot(realisations,
        label="Actual values ($x_t$)")
    plt.legend(loc="upper left")
    plt.grid(True)
    f.show()
```

Listing 60 Plotting Simple Exponential Smoothing forecasts in Python.

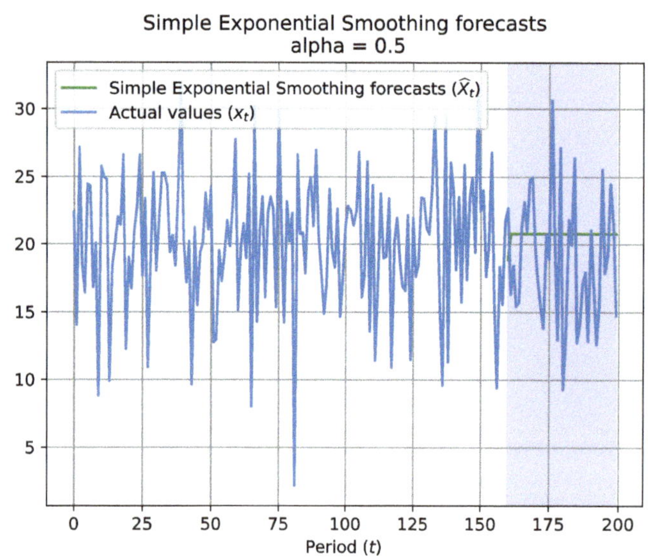

Fig. 72 Forecasts for the last 40 periods by using Simple Exponential Smoothing with a smoothing parameter $\alpha = 0.5$; the underpinning stochastic process is a a Gaussian process with mean $\mu = 20$ and standard deviation $\sigma = 5$.

Simple Exponential Smoothing method one-step forecasts can be computed as follows.

```
def ses_rolling(series, alpha, x0):
   forecasts = np.empty(len(series))
   forecasts[0] = x0
   for k in range(1,len(series)):
      forecasts[k] = alpha*series[k-1] + (1-alpha)*forecasts[k-1]
   return forecasts
```

By leveraging these forecasts, we carry out residuals analysis.

```
N, t, alpha, x0 = 200, 160, 0.5, 20
realisations = pd.Series(list(sample_random_walk(0, N)), range(N))
forecasts = ses_rolling(realisations, alpha, x0)
residuals = residuals(realisations, forecasts)
print("E[e_t] = "+str(statistics.mean(residuals)))
standardised_residuals = standardised_residuals(realisations, forecasts)
residuals_plot(residuals)
residuals_histogram(standardised_residuals)
residuals_autocorrelation(residuals, None)
sm.qqplot(standardised_residuals, line ='45')
py.show()
```

Residuals have mean -0.018, which is close to zero (Fig. 73).

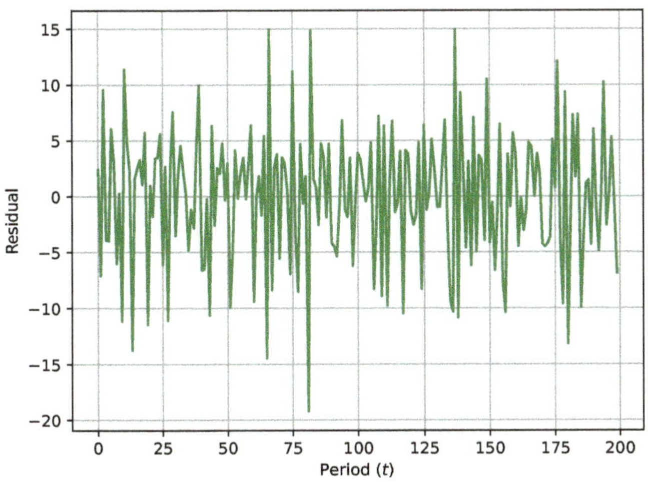

Fig. 73 Residual analysis for Simple Exponential Smoothing: residuals.

The histogram in Fig. 51 suggests that residuals are approximately Gaussian. Fig. 52 reveals absence of residuals autocorrelation. Finally, the Q-Q plot (Fig. 53) appears to further support normality of residuals.

Instead of reimplementing Simple Exponential Smoothing from scratch, we can rely on Python library `statsmodels.tsa.api`, which provides readily available apis for time series analysis.

```
from statsmodels.tsa.api import SimpleExpSmoothing

N, t, alpha, x0 = 200, 160, 0.5, 20
realisations = pd.Series(sample_gaussian_process(20, 5, N), range(N))
mod = SimpleExpSmoothing(realisations[:t+1]).fit(smoothing_level=alpha,
      initial_level=x0, optimized=False)
forecasts = mod.forecast(N-(t+1)).rename(r'$\alpha=0.5$')
plot(realisations, pd.Series(np.nan, range(t+1)).append(forecasts), alpha)
py.show()
```

This code produces exactly the same results illustrated in Fig. 72.

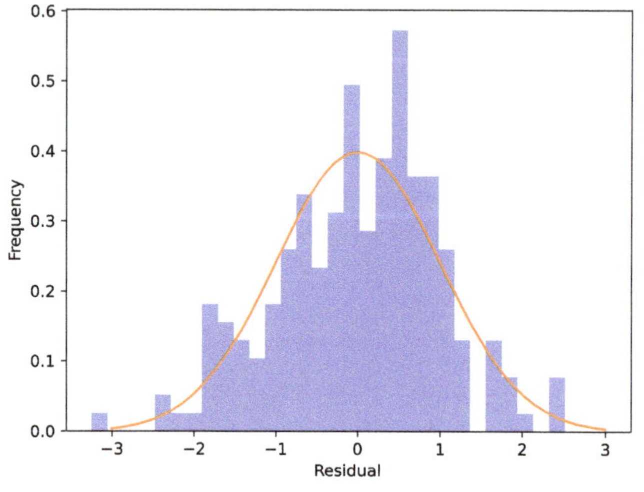

Fig. 74 Residual analysis for Simple Exponential Smoothing: histogram.

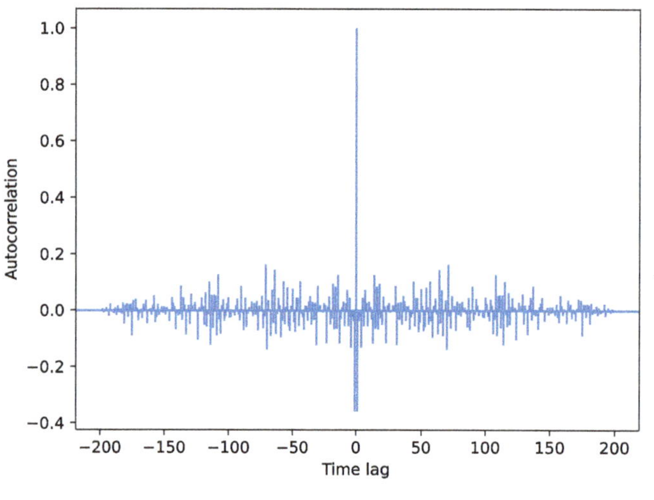

Fig. 75 Residual analysis for Simple Exponential Smoothing: autocorrelation plot.

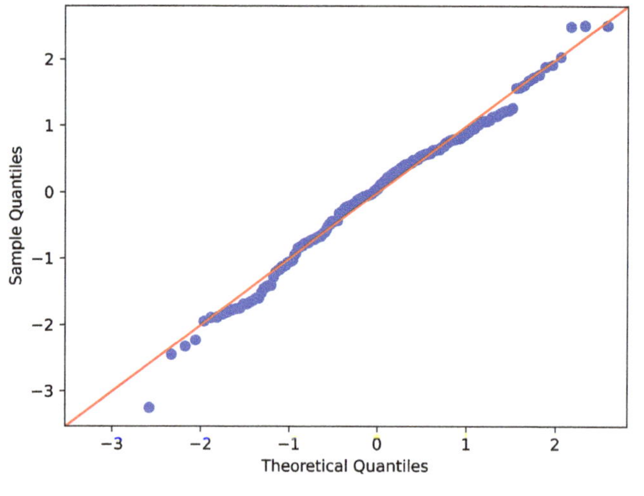

Fig. 76 Residual analysis for Simple Exponential Smoothing: Q-Q plot.

By relying on a state space formulation,[36] implemented in package statsmodels.tsa.statespace.exponential_smoothing, we derive prediction intervals (Fig. 77).

[36] James Durbin and Siem Jan Koopman. *Time series analysis by state space methods*. Oxford Univ. Pr., 2001.

```
from statsmodels.tsa.statespace.exponential_smoothing import ExponentialSmoothing

N, t, alpha, x0 = 200, 160, 0.5, 20
realisations = pd.Series(sample_gaussian_process(20, 5, N), range(N))
mod = ExponentialSmoothing(realisations[:t+1], initialization_method='known',
        initial_level=x0).fit(disp=False)
print(mod.summary())
forecasts = mod.get_forecast(N-(t+1))
forecasts_ci = forecasts.conf_int(alpha=0.05)
plot_ci(realisations, pd.Series(np.nan,
        range(t+1)).append(forecasts.predicted_mean), forecasts_ci, alpha)
py.show()
```

```
def plot_ci(realisations, forecasts,
        forecasts_ci, alpha):
    f = plt.figure(1)
    plt.title("Simple Exponential
            Smoothing forecasts\n State
            Space Model")
    plt.xlabel('Period ($t$)')
    first, last = next(x for x, val in
            enumerate(forecasts) if
            ~np.isnan(val)),
            len(forecasts)-1
    plt.axvspan(first, last, alpha=0.2,
            color='blue')
    plt.plot(forecasts, "g",
            label="Simple Exponential
            Smoothing forecasts
            ($\widehat{X}_t$)")
    plt.plot(realisations,
            label="Actual values ($x_t$)")
    t = next(x for x, val in
            enumerate(forecasts) if
            ~np.isnan(val)) - 1
    forecast_index = np.arange(t+1, t+1
            + len(forecasts_ci))
    plt.fill_between(forecast_index,
            forecasts_ci.iloc[:, 0],
            forecasts_ci.iloc[:, 1],
            color='r', alpha=0.1)
    plt.legend(loc="upper left")
    plt.grid(True)
    f.show()
```

Listing 61 Plotting Simple Exponential Smoothing forecasts and prediction intervals in Python.

Listing 61 illustrates the plot function.

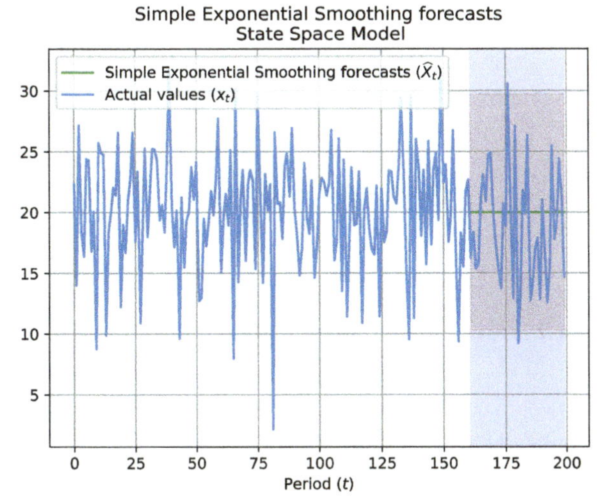

Fig. 77 Simple Exponential Smoothing forecasts and prediction intervals for the last 40 periods of a Gaussian process with mean $\mu = 20$ and standard deviation $\sigma = 5$.

In Fig. 78 we illustrate Simple Exponential Smoothing forecasts and prediction intervals for a random walk with standard Gaussian noise; these are similar to those obtained with the Naïve method.

```
realisations = pd.Series(list(sample_random_walk(0, N)), range(N))
```

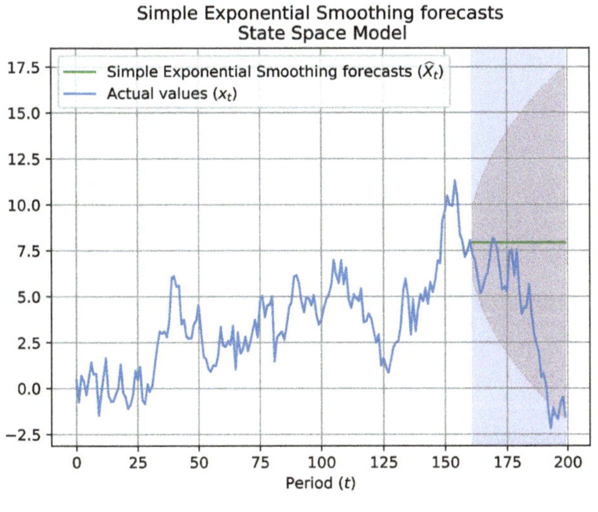

Fig. 78 Simple Exponential Smoothing forecasts and prediction intervals for the last 40 periods of a random walk with standard Gaussian noise.

Double Exponential Smoothing (Holt's method)

As previously discussed, Simple Exponential Smoothing has a "flat" forecast function: all future forecasts take the same value, which represents the "the level" (or the smoothed value) of the series. This means the method will only be suitable if the time series has *no trend or seasonal component*.

We shall next focus on time series featuring a linear trend, such as the random walk with drift previously considered in the context of the Drift method.

Holt extended[37] Simple Exponential Smoothing to allow the forecasting of a time series with a trend. In Holt's method, given past realisations $\{x_1, x_2, \ldots, x_t\}$, the forecast is defined as

$$\widehat{X}_{t+k} \triangleq l_t + kb_t$$

where l_t denotes an estimate of the level of the series at time t, and b_t denotes an estimate of the trend (slope) of the series at time t. These level and trend estimates are obtained by means of the following *smoothing equations*

$$l_t = \alpha x_t + (1-\alpha)(l_{t-1} + b_{t-1}) \qquad \text{(level equation)}$$
$$b_t = \beta(l_t - l_{t-1}) + (1-\beta)b_{t-1} \qquad \text{(trend equation)}$$

where $0 < \alpha < 1$ and $0 < \beta < 1$ are the smoothing parameters for the level and trend, respectively. Holt's method is available in library `statsmodels.tsa.api` and can be implemented as follows

```
from statsmodels.tsa.api import Holt

N, t = 200, 160
realisations = pd.Series(list(sample_random_walk(0, 0.1, N)), range(N))
mod = Holt(realisations[:t+1]).fit(optimized=True)
params = ['smoothing_level', 'smoothing_trend', 'initial_level', 'initial_trend']
results=pd.DataFrame(index=["alpha","beta","l_0","b_0","SSE"] ,columns=["Holt's"])
results["Holt's"] = [mod.params[p] for p in params] + [mod.sse]
print(results)
forecasts = mod.forecast(N-(t+1)).rename(r'$\alpha=0.5$ and $\beta=0.5$')
plot(realisations, pd.Series(np.nan, range(t+1)).append(forecasts))
plot_components(mod)
py.show()
```

where `sample_random_walk` is the function presented in Listing 55. Note that in some version of the library `smoothing_trend` and `initial_trend` become `smoothing_slope` and `initial_slope`. Listing 62 illustrates the `plot` and `plot_components` functions.

Whilst it is possible to manually set values of model parameters, function `fit` also allows to automatically estimate (`optimized=True`) model parameters such as the initial level (l_0) and the initial trend (b_0), as well as the two smoothing parameters α and β. Parameters automatically estimated by the function are shown in Table 9.

The level and slope components resulting from Holt's decomposition are shown in Fig. 79, which has been generated by function `plot_components`.

Finally, prediction intervals can be obtained once more by leveraging the state space formulation implemented in package `statsmodels.tsa.statespace.exponential_smoothing`.

[37] Charles C. Holt. Forecasting seasonals and trends by exponentially weighted moving averages. *International Journal of Forecasting*, 20(1):5–10, 2004.

```
def plot(realisations, forecasts):
    f = plt.figure(1)
    plt.title("Holt's forecasts")
    plt.xlabel('Period ($t$)')
    first, last = next(x for x, val in
        enumerate(forecasts) if
        ~np.isnan(val)),
        len(forecasts)-1
    plt.axvspan(first, last, alpha=0.2,
        color='blue')
    plt.plot(forecasts, "g",
        label="Holt's forecasts
        ($\widehat{X}_t$)")
    plt.plot(realisations,
        label="Actual values ($x_t$)")
    plt.legend(loc="upper left")
    plt.grid(True)
    f.show()

def plot_components(fit):
    f = plt.figure(1)
    pd.DataFrame(np.c_[fit.level,
        fit.trend]).rename(
      columns={0:'level',
          1:'trend'}).plot(
      subplots=True)
    plt.xlabel('Period ($t$)')
    f.show()
```

Listing 62 Plotting Holt's method forecasts and components in Python.

	Holt's
α	0.797
β	0.000
l_0	0.225
b_0	0.148
SSE	151

Table 8 Holt's method fitted model parameters and Sum of Squared Errors (SSE).

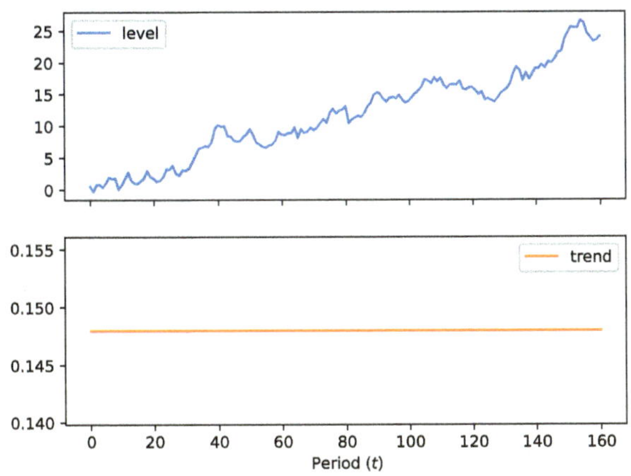

Fig. 79 Holt's method: level and slope components of a random walk with standard Gaussian noise and drift $c = 0.1$.

```
def plot_ci(realisations, forecasts, forecasts_ci):
    f = plt.figure(1)
    plt.title("Holt's forecasts\n State Space Model")
    plt.xlabel('Period ($t$)')
    first, last = next(x for x, val in enumerate(forecasts) if ~np.isnan(val)),
        len(forecasts)-1
    plt.axvspan(first, last, alpha=0.2, color='blue')
    plt.plot(forecasts, "g", label="Holt's forecasts ($\widehat{X}_t$)")
    plt.plot(realisations, label="Actual values ($x_t$)")
    t = next(x for x, val in enumerate(forecasts) if ~np.isnan(val)) - 1
    forecast_index = np.arange(t+1, t+1 + len(forecasts_ci))
    plt.fill_between(forecast_index, forecasts_ci.iloc[:, 0], forecasts_ci.iloc[:,
        1], color='r', alpha=0.1)
    plt.legend(loc="upper left")
    plt.grid(True)
    f.show()

N, t = 200, 160
realisations = pd.Series(list(sample_random_walk(0, 0.1, N)), range(N))
mod = ExponentialSmoothing(realisations[:t+1], trend=True,
      initialization_method='estimated').fit(disp=False)
print(mod.summary())
forecasts = mod.get_forecast(N-(t+1))
forecasts_ci = forecasts.conf_int(alpha=0.05)
plot_ci(realisations, pd.Series(np.nan,
      range(t+1)).append(forecasts.predicted_mean), forecasts_ci)
py.show()
```

The results are illustrated in Fig. 80.

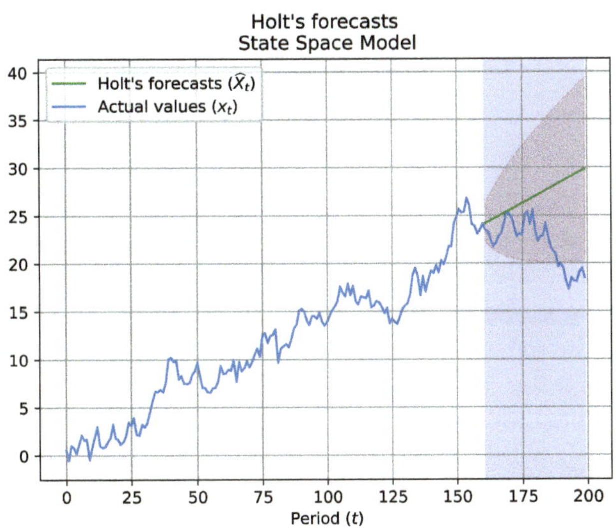

Fig. 80 Holt's method forecasts and prediction intervals for a random walk with standard Gaussian noise and drift $c = 0.1$.

Triple Exponential Smoothing (Holt-Winters' seasonal method)

Holt's method is not suitable if the time series features a *seasonal component*. We shall here focus on time series featuring such component. One of such series is the seasonal random walk previously considered in the context of the Seasonal Naïve method.

Holt[38] and Winters[39] extended Holt's method to capture a seasonal component. They discussed both additive and multiplicative variants of their method. For the sake of brevity, we shall limit our discussion to the additive method. In Holt-Winters' method, given past realisations $\{x_1, x_2, \ldots, x_t\}$, the forecast is defined as

$$\widehat{X}_{t+k} \triangleq l_t + kb_t + s_{t+k-m\lfloor (k-1)/m \rfloor}$$

where $\lfloor x \rfloor$ denotes the integer part of x, l_t denotes an estimate of the level of the series at time t, b_t denotes an estimate of the trend (slope) of the series at time t, and s_t denotes an estimate of the seasonal component of the series at time t. We use m to denote the frequency of the seasonality, that is the number of seasons in a year. For example, for quarterly data $m = 4$, for monthly data $m = 12$.

Level, trend, and seasonal component estimates are obtained by means of the following *smoothing equations*

$$l_t = \alpha(x_t - s_{t-m}) + (1-\alpha)(l_{t-1} + b_{t-1}) \quad \text{(level equation)}$$
$$b_t = \beta(l_t - l_{t-1}) + (1-\beta)b_{t-1} \quad \text{(trend equation)}$$
$$s_t = \gamma(x_t - l_t) + (1-\gamma)s_{t-m} \quad \text{(seasonal equation)}$$

where $0 < \alpha < 1, 0 < \beta < 1, 0 < \gamma < 1$ are the smoothing parameters for the level, trend, and seasonal component, respectively. Holt-Winters' method method is available in library `statsmodels.tsa.api` and can be implemented as follows

```
from statsmodels.tsa.api import ExponentialSmoothing

N, t, m = 100, 80, 4
realisations = pd.Series(list(sample_seasonal_random_walk(N,m)), range(N))
mod = ExponentialSmoothing(realisations[:t+1], seasonal_periods=4, trend='add',
      seasonal='add').fit(optimized=True)
params = ['smoothing_level', 'smoothing_trend', 'smoothing_seasonal',
      'initial_level', 'initial_trend']
results=pd.DataFrame(index=["alpha","beta","gamma","l_0","b_0","SSE"]
      ,columns=["Holt-Winters'"])
results["Holt-Winters'"] = [mod.params[p] for p in params] + [mod.sse]
print(results)
forecasts = mod.forecast(N-(t+1)).rename(r'$\alpha=0.5$ and $\beta=0.5$')
plot(realisations, pd.Series(np.nan, range(t+1)).append(forecasts))
plot_components(mod)
py.show()
```

where `sample_seasonal_random_walk` is the function presented in Listing 57. listing 63 illustrates the `plot` and `plot_components` functions. Whilst it is possible to manually set values of model parameters, function `fit` also allows to automatically estimate (`optimized=True`) model parameters such as the initial level (l_0) and the initial trend (b_0), as well as the three smoothing parameters α, β, and γ. Parameters automatically estimated by the function are shown in Table 9. The level and slope components resulting from Holt-Winters' decomposition are shown in Fig. 81, which has been generated by function `plot_components`.

[38] Charles C. Holt. Forecasting seasonals and trends by exponentially weighted moving averages. *International Journal of Forecasting*, 20(1):5–10, 2004.

[39] Peter R. Winters. Forecasting sales by exponentially weighted moving averages. *Management Science*, 6(3): 324–342, 1960.

```
def plot(realisations, forecasts):
    f = plt.figure(1)
    plt.title("Holt-Winters' forecasts")
    plt.xlabel('Period ($t$)')
    first, last = next(x for x, val in
        enumerate(forecasts) if
        ~np.isnan(val)),
        len(forecasts)-1
    plt.axvspan(first, last, alpha=0.2,
        color='blue')
    plt.plot(realisations,
        label="Actual values ($x_t$)")
    plt.plot(forecasts, "g",
        label="Holt-Winters'
        forecasts ($\widehat{X}_t$)")
    plt.legend(loc="upper left")
    plt.grid(True)
    f.show()

def plot_components(fit):
    f = plt.figure(1)
    pd.DataFrame(np.c_[fit.level,
        fit.trend,
        fit.season]).rename(
      columns={0:'level', 1:'trend',
          2:'seasonal'}).plot(
          subplots=True)
    plt.xlabel('Period ($t$)')
    f.show()
```

Listing 63 Plotting Holt-Winters' method forecasts and components in Python.

	Holt-Winters'
α	0.000
β	0.000
γ	0.839
l_0	2.48
b_0	0.015
SSE	71.2

Table 9 Holt-Winters' method fitted model parameters and Sum of Squared Errors (SSE).

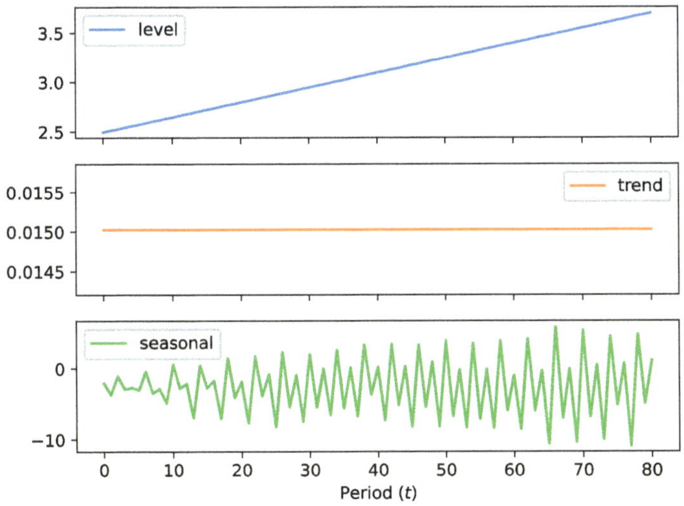

Fig. 81 Holt-Winters' method: level, slope, and seasonal components of a seasonal random walk with standard Gaussian noise and $m = 4$.

```
def plot_ci(realisations, forecasts,
      forecasts_ci):
    f = plt.figure(1)
    plt.title("Holt-Winters'
          forecasts\n State Space
          Model")
    plt.xlabel('Period ($t$)')
    first, last = next(x for x, val in
          enumerate(forecasts) if
          ~np.isnan(val)),
          len(forecasts)-1
    plt.axvspan(first, last, alpha=0.2,
          color='blue')
    plt.plot(realisations,
          label="Actual values ($x_t$)")
    plt.plot(forecasts, "g",
          label="Holt-Winters'
          forecasts ($\widehat{X}_t$)")
    t = next(x for x, val in
          enumerate(forecasts) if
          ~np.isnan(val)) - 1
    forecast_index = np.arange(t+1, t+1
          + len(forecasts_ci))
    plt.fill_between(forecast_index,
          forecasts_ci.iloc[:, 0],
          forecasts_ci.iloc[:, 1],
          color='r', alpha=0.2)
    plt.legend(loc="upper left")
    plt.grid(True)
    f.show()
```

Listing 64 Plotting Holt-Winters' method forecasts and prediction intervals in Python.

Finally, prediction intervals can be obtained once more by leveraging the state space formulation implemented in package `statsmodels.tsa.statespace.exponential_smoothing` (Fig. 82).

```
from statsmodels.tsa.statespace.exponential_smoothing import ExponentialSmoothing

N, t, m = 100, 80, 4
realisations = pd.Series(list(sample_seasonal_random_walk(N,m)), range(N))
mod = ExponentialSmoothing(realisations[:t+1], trend='add', seasonal='add',
      initialization_method='estimated').fit(disp=False)
print(mod.summary())
forecasts = mod.get_forecast(N-(t+1))
forecasts_ci = forecasts.conf_int(alpha=0.05)
plot_ci(realisations, pd.Series(np.nan,
      range(t+1)).append(forecasts.predicted_mean), forecasts_ci)
py.show()
```

Listing 64 illustrates the `plot` function.

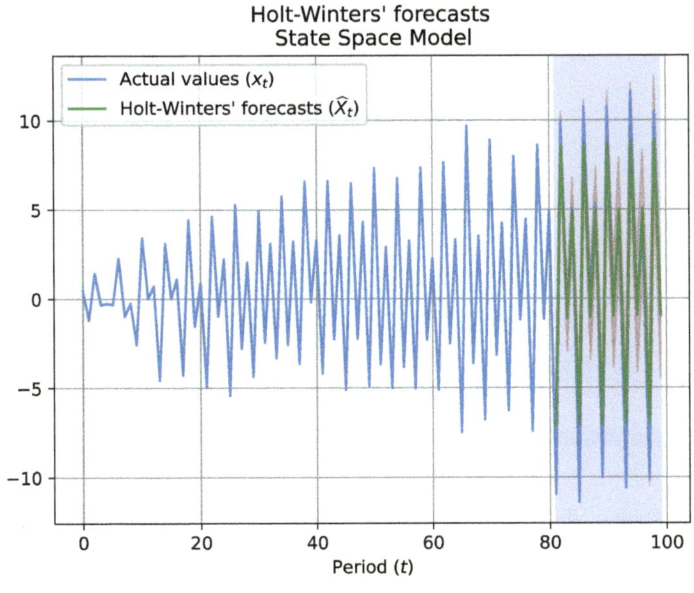

Fig. 82 Holt-Winters' method forecasts and prediction intervals for a seasonal random walk with standard Gaussian noise and $m = 4$.

ARIMA models

In practice, it is often the case that stochastic demands in different periods are correlated; for instance, this may happen when we only serve a few large customers, so that if demand is high at a given period, one may expect demand in the following periods to be lower (i.e. negatively correlated), because a high demand may indicate that several customers have replenished their stock. Autocorrelation, also known as serial correlation, is the correlation of a signal with a delayed copy of itself as a function of delay. Forecasting techniques that aim to describe autocorrelation in the data have been developed by Box and Jenkins.[40]

[40] George. E. P. Box and Gwilym M. Jenkins. *Time series analysis: Forecasting and control.* Holden-Day, 1976.

Differencing

We have seen in the section illustrating Box-Cox transformations that logarithms and power transformations can help stabilising the variance of a time series.

Conversely, *differencing can help stabilise the mean* of a time series by removing changes in the level of a time series, and therefore eliminating (or reducing) trend and seasonality.

A differenced time series y'_t is the change between consecutive observations in the original series, and can be written as

$$y'_t = y_t - y_{t-1}.$$

When the differenced series is a white noise, the model for the original series can be written as

$$y_t - y_{t-1} = \varepsilon_t,$$

where ε_t is a white noise. By rearranging, we obtain $y_t = y_{t-1} + \varepsilon_t$, which suggests that the series is a realisation of a random walk. If the differences have non-zero mean, say c, the series can be expressed as $y_t = c + y_{t-1} + \varepsilon_t$, that is as a random walk with drift.

A seasonal difference (or "lag-m difference") is the difference between an observation and the previous observation from the same season. If seasonally differenced data appear to be white noise, the series is a realisation of a seasonal random walk (Fig. 83).

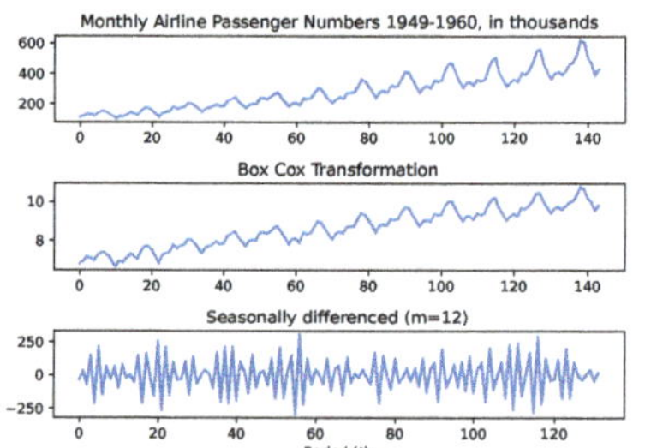

Fig. 83 Airline time series: Box-Cox transformation and seasonal differencing.

```
import statsmodels.api as sm, pandas as pd
import matplotlib.pyplot as plt
import scipy.stats as stats
from statsmodels.tsa.statespace.tools import diff

airpass = sm.datasets.get_rdataset("AirPassengers", "datasets")
fig, axs = plt.subplots(3)
axs[0].set_title('Monthly Airline Passenger Numbers 1949-1960, in thousands')
axs[0].plot(pd.Series(airpass.data["value"]))
series, l = stats.boxcox(airpass.data["value"])
axs[1].plot(series)
axs[1].set_title('Box Cox Transformation')
differenced = diff(series, k_diff=12)
axs[2].plot(differenced)
axs[2].set_title('Seasonally differenced (m=12)')
plt.xlabel('Period ($t$)')
fig.tight_layout()
plt.show()
```

Autoregressive (AR) model

In an autoregression model, we forecast the variable of interest using a linear combination of past values of the variable. The term autoregression indicates that it is a regression of the variable against itself.

Definition 15. *An autoregressive model of order p, AR(p) in short, is defined as*

$$X_t \triangleq c + \sum_{i=1}^{p} \varphi_i X_{t-i} + \varepsilon_t,$$

where $\varphi_1, \ldots, \varphi_p$ are the parameters of the model, c is a constant, and ε_t is a white noise.

$\varphi_1 = 0$	$c = 0$	white noise
$\varphi_1 = 1$	$c = 0$	random walk
$\varphi_1 = 1$	$c \neq 0$	random walk with drift

Table 10 Special cases of AR(1).

Note that the variance of the error term ε_t only affects the scale of the series, not the patterns. By varying the parameters of the model, we can obtain a wide range of different time series patterns, some of which have been surveyed before (Table 10).

It is common to apply autoregressive models under the assumption that the model underpinning the time series is stationary; for this reason, some constraints on the values of the parameters are required (Table 11). Typically, these constraints are automatically enforced when a model is fit by an off-the-shelf software package.

AR(1)	$\|\varphi_1\| < 1$
AR(2)	$\|\varphi_1\| < 1$, $\|\varphi_2\| < 1$, $\|\varphi_1 + \varphi_2\| < 1$, $\|\varphi_2 - \varphi_1\| < 1$
AR(p)	roots of $1 - \sum_{i=1}^{p} \varphi_i z^{p-i}$ must lie outside the unit circle.

Table 11 Restrictions to model parameters that ensures stationarity.

Example 26. *Let $\{X_t\}$ be a random walk with standard Gaussian noise $\{\varepsilon_t\}$. We sample 200 realisations from this stochastic process as previously shown in Listing 53. By leveraging* `statsmodels.tsa.ar_model.AutoReg` *we fit an AR(1) model as follows.*

```
import numpy as np, pandas as pd, statistics
from statsmodels.tsa.ar_model import AutoReg

N, t, p = 200, 180, 1
realisations = pd.Series(list(sample_random_walk(0, N)), range(N))
mod = AutoReg(realisations[0:t], p)
res = mod.fit()
print(res.summary())
print("Std residuals: "+str(statistics.stdev(res.resid)))
```

The result of the fitting procedure is the following.

```
                            AutoReg Model Results
==============================================================================
Dep. Variable:                      y   No. Observations:                  180
Model:                     AutoReg(1)   Log Likelihood                -248.301
Method:               Conditional MLE   S.D. of innovations              0.969
Date:                Mon, 22 Feb 2021   AIC                             -0.030
Time:                        00:12:02   BIC                              0.023
Sample:                             1   HQIC                            -0.008
                                  180
==============================================================================
                 coef    std err          z      P>|z|      [0.025      0.975]
------------------------------------------------------------------------------
intercept      0.2950      0.126      2.340      0.019       0.048       0.542
y.L1           0.9327      0.027     34.319      0.000       0.879       0.986
                                    Roots
=============================================================================
                  Real          Imaginary           Modulus         Frequency
-----------------------------------------------------------------------------
AR.1            1.0721           +0.0000j            1.0721            0.0000
-----------------------------------------------------------------------------
Std residuals: 0.9714322305049528
```

`statsmodels.tsa.ar_model.ar_select_order` automatically selects the order p of an AR(p) process that best fits the data.

```
from statsmodels.tsa.ar_model import ar_select_order

N, t, p, max_order = 200, 180, 1, 10
realisations = pd.Series(list(sample_random_walk(0, N)), range(N))
sel = ar_select_order(realisations[0:t], max_order)
res = sel.model.fit()
print(res.summary())
print("Std residuals: "+str(statistics.stdev(res.resid)))
```

In our example this leads to the same result ($p = 1$), that is an AR(1) process. Fitting diagnostics can be obtained as follows.

```
import matplotlib.pyplot as plt

fig = plt.figure(figsize=(16,9))
res.plot_diagnostics(fig=fig, lags=max_order)
plt.show()
```

The diagnostics for our example are shown in Fig. 84. The plots illustrating standardised residuals, residual distribution histogram, and Q-Q plot are similar to those previously presented. However, the diagnostics here presented also include a "correlogram."[41] In this case, we see a spike at value 1 in the x-axis (which represents the order p of the process), while values for $x = 2, 3, \ldots$ appear to be random fluctuations that remain inside the confidence bands illustrated and are not significantly different than 0. This suggests that there is evidence of correlation between a period and the previous one, as it is effectively the case in a random walk.

[41] A correlogram (also called Auto Correlation Function ACF Plot or Autocorrelation plot) is a visual way to show serial correlation in data that changes over time.

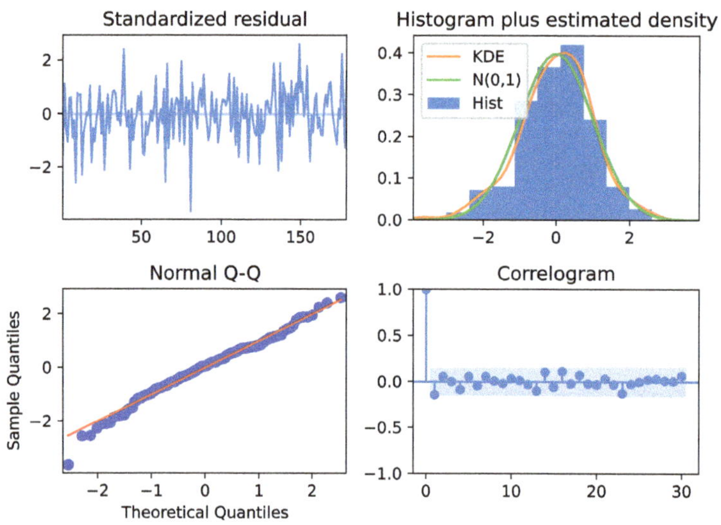

Fig. 84 Fitting an AR(1) to a random walk: diagnostics.

Finally, one can produce forecasts and prediction intervals by using `plot_predict` (Fig. 85). The forecasts and prediction intervals obtained are similar to those produced by the Naïve method for the same example (Fig. 69).

```
res.plot_predict(start=t, end=N)
plt.plot(realisations[0:N], label="realisations")
plt.legend(loc="upper left")
plt.grid(True)
plt.show()
```

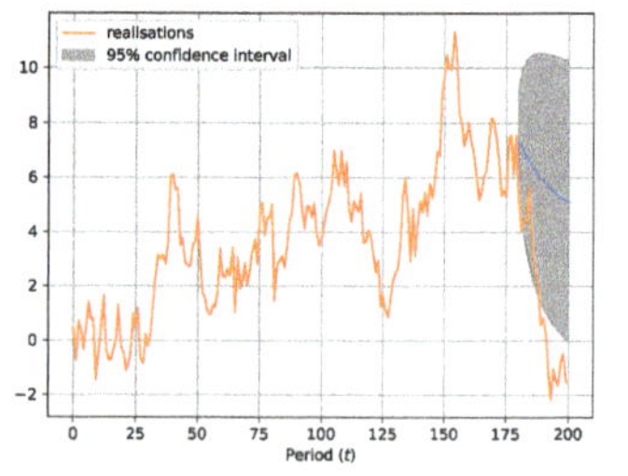

Fig. 85 Forecasts for periods $180, \ldots, 200$ for an AR(1) process fit to a random walk.

Moving Average (MA) model

The Moving Average model specifies that the output variable depends linearly on the current and various past values of a stochastic error $\{\varepsilon_t\}$.

Definition 16. *A Moving Average model of order q, MA(q) in short, is defined as*

$$X_t \triangleq \mu + \varepsilon_t + \theta_1 \varepsilon_{t-1} + \ldots + \theta_q \varepsilon_{t-q},$$

where μ is the mean of the series, $\theta_1, \ldots, \theta_q$ are the parameters of the model, and $\{\varepsilon_t\}$ is a white noise.

In this model, X_t can be thought of as a weighted moving average of the past few forecast errors.

Lemma 33. *The finite MA model is always stationary.*

Lemma 34. *Any stationary AR(p) model can be written as an MA(∞) model.*

Definition 17. *An MA(q) model is invertible if it can be expressed as an AR(∞) model.*

For an MA(1) it is easy to show that the model is invertible if and only if $|\theta| < 1$; in fact, $|\theta| \geq 1$ means that more distant observations have greater or equal influence on the current error than closer ones — a situation that does not make much sense. Constraints can be imposed on MA(q) model to ensure invertibility; these are automatically enforced when a model is fit by a software package.

To sample from an MA(q) process we can leverage Listing 65. Alternatively, `statsmodels.tsa.arima_process.ArmaProcess` combines an AR(p) and an MA(q) to obtain a so-called ARMA(p, q) process. By setting $p = 0$, an ARMA(p, q) process reduces to an MA(q) process, which can be sampled (Listing 66).

```
def sample_MA_process(mu, theta,
      realisations):
    np.random.seed(1234)
    errors = np.random.normal(0, 1,
        realisations + len(theta))
    theta = np.r_[1, theta][::-1]
    for r in range(1,
        min(len(theta),realisations+1)):
      yield mu +
          sum(np.multiply(theta[-r:],
          errors[:r]))
    for r in
        range(realisations-len(theta)+1):
      yield mu +
          sum(np.multiply(theta,
          errors[r:r+len(theta)]))
```

Listing 65 Sampling an MA(q) process.

Example 27. *Let $\{X_t\}$ be an MA(q) process subject to standard Gaussian noise $\{\varepsilon_t\}$, $q = 2$, and parameters $\theta_1 = 0.8$ and $\theta_2 = 0.2$. We sample 200 realisations from this stochastic process.*

```
from statsmodels.tsa.arima_process
     import ArmaProcess

def sample_ARMA0q_process(mu, theta,
     realisations):
   np.random.seed(1234)
   dist = lambda size:
       np.random.normal(0, 1, size)
   arparams = np.array([])
   maparams = np.array(theta)
   # include zero-th lag
   arparams = np.r_[1, arparams]
   maparams = np.r_[1, maparams]
   arma_t = ArmaProcess(arparams,
       maparams)
   return
       arma_t.generate_sample(nsample
       = realisations, distrvs=dist)
```

Listing 66 Sampling an ARMA($0, q$) process.

By using `statsmodels.api.tsa.ARMA` we fit an MA(q) model.

```
import numpy as np, pandas as pd, statsmodels.api as sm, statistics

mu, theta, N, t, max_order = 0, [0.8,0.2], 200, 180, 10
realisations = pd.Series(list(sample_MA_process_ARMA(mu, theta, N)), range(N))
mod = sm.tsa.ARMA(realisations[0:t], order=(0, 2))
res = mod.fit()
print(res.summary())
print("Std residuals: "+str(statistics.stdev(res.resid)))
```

The result after fitting an MA(2) model is shown below.

```
                              ARMA Model Results
==============================================================================
Dep. Variable:                      y   No. Observations:                  180
Model:                     ARMA(0, 2)   Log Likelihood                -249.163
Method:                       css-mle   S.D. of innovations              0.965
Date:                Mon, 22 Feb 2021   AIC                            506.326
Time:                        00:27:11   BIC                            519.098
Sample:                             0   HQIC                           511.505
==============================================================================
                 coef    std err          z      P>|z|      [0.025      0.975]
------------------------------------------------------------------------------
const          0.0847      0.122      0.693      0.488      -0.155       0.324
ma.L1.y        0.6220      0.074      8.448      0.000       0.478       0.766
ma.L2.y        0.0820      0.081      1.015      0.310      -0.076       0.241
                                    Roots
=============================================================================
                  Real          Imaginary           Modulus         Frequency
-----------------------------------------------------------------------------
MA.1           -2.3137           +0.0000j            2.3137            0.5000
MA.2           -5.2682           +0.0000j            5.2682            0.5000
-----------------------------------------------------------------------------
Std residuals: 0.967862299006029
```

While fitting the MA(q) model to the data, we assumed the order q of the process was known. If this is not the case, one may analyse the autocorrelation of the data. The autocorrelation function (ACF) of an MA(q) process is zero at lag $q+1$ and greater, thus it easy to determine the correct value of q by inspecting the ACF (Fig. 86).

```
sm.graphics.tsa.plot_acf(realisations.values.squeeze(), lags=max_order)
```

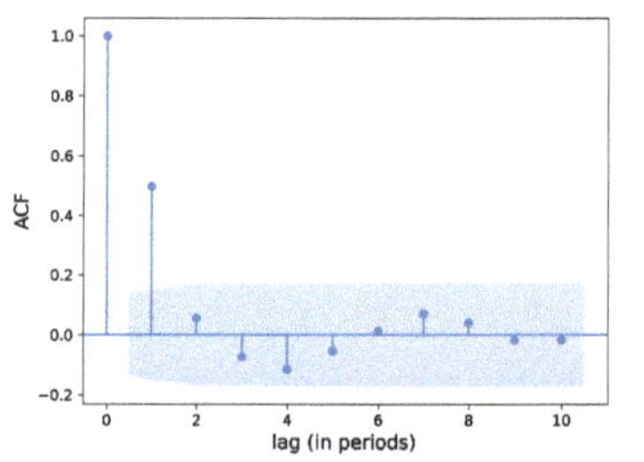

Fig. 86 Fitting an MA(2) model: inspecting the ACF. In this instance the ACF is nonzero for the first two lags.

The order of an ARMA(p,q) process can also be estimated via the following function.

```
from statsmodels.tsa.stattools import arma_order_select_ic

order = arma_order_select_ic(realisations[0:t], ic='aic', max_ar = 5, max_ma = 5)
print(order.aic_min_order)
```

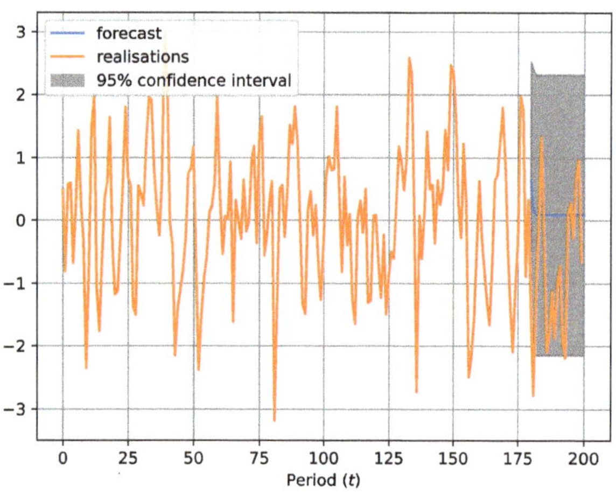

Fig. 87 Forecasts for periods $180, \ldots, 200$ for an MA(2) process.

Finally, one can produce forecasts and prediction intervals by using `plot_predict` (Fig. 87).

```
import matplotlib.pyplot as plt

res.plot_predict(start=t, end=N, plot_insample=False)
plt.plot(realisations[0:N], label="realisations")
plt.legend(loc="upper left")
plt.grid(True)
plt.show()
```

Autoregressive Integrated Moving Average (ARIMA) model

ARIMA is an acronym for AutoRegressive Integrated Moving Average (in this context, "integration" is the reverse of differencing). The I (for "integrated") indicates that the data values have been replaced with the difference between their values and the previous values, and this differencing process may have been performed more than once.

Definition 18. *An ARIMA(p, d, q) model can be written as*

$$X'_t \triangleq c + \varphi_1 y'_{t-1} + \ldots + \varphi_p y'_{t-p} + \theta_1 \varepsilon_{t-1} + \ldots + \theta_q \varepsilon_{t-q} + \varepsilon_t,$$

where c is a constant; X'_t is the differenced stochastic process (note that the process may have been differenced more than once); and $\{\varepsilon_t\}$ is a white noise. In an ARIMA(p, d, q) we define

p	*order of the autoregressive part;*
d	*degree of first differencing involved;*
q	*order of the moving average part.*

The same stationarity and invertibility conditions that are used for AR and MA models also apply to an ARIMA model.

Many of the models we have previously considered are special cases of the ARIMA model, in particular

White noise	ARIMA(0, 0, 0)
Random walk	ARIMA(0, 1, 0) with no constant
Random walk with drift	ARIMA(0, 1, 0) with a constant
Autoregression	ARIMA(p, 0, 0)
Moving average	ARIMA(0, 0, q)

Most software packages for time series analysis provide facilities to automatically determine values of p, d, and q.

Example 28. *We fit an ARIMA(0, 1, 0) to a time series reporting fluctuations in air passenger numbers (Listing 67), and leverage the fitted model to produce forecasts and prediction intervals (Fig. 88).*

```
import statsmodels.api as sm, pandas
    as pd, statistics
from statsmodels.tsa.arima_model
    import ARIMA
import matplotlib.pyplot as plt

N, t = 140, 136
airpass = sm.datasets.get_rdataset(
    "AirPassengers", "datasets")
ts = pd.Series(airpass.data["value"])
ts = ts.astype(float)
model = ARIMA(ts[0:t], order=(0,1,0))
res = model.fit()
print(res.summary())
res.plot_predict(start=ts.index[3],
    end=ts.index[-1], alpha=0.1)
plt.xlabel('Period ($t$)')
print("Std residuals:
    "+str(statistics.stdev(res.resid)))
plt.show()
```

Listing 67 Fitting an ARIMA(0, 1, 0) to a time series reporting fluctuations in air passenger numbers.

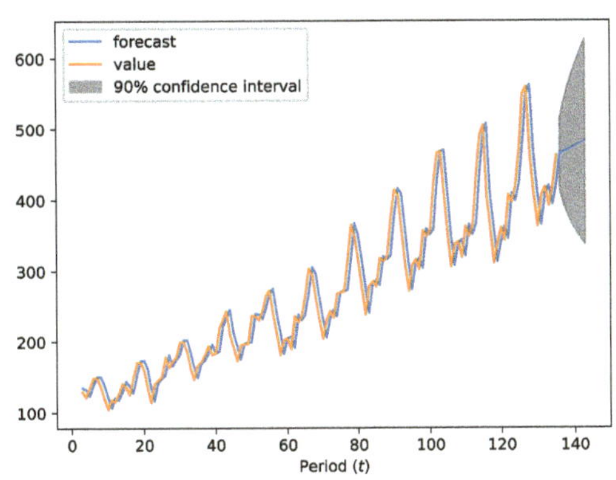

Fig. 88 Airline time series: forecasts and prediction intervals obtained by fitting an ARIMA(0, 1, 0) model. The fitted constant is $c = 2.58$; the standard deviation of residuals is $\sigma = 31.2$.

Practical considerations

We conclude this chapter by emphasising the "ancillary" nature of the predictive analysis carried out in this chapter. To be precise, any predictive analysis is always ancillary to an associated prescriptive analysis, which must necessarily be present, as

> the only reason why we make predictions, is to act upon them.

While one may invest considerable time — and sometime have fun — fiddling with models and data, contrasting forecast quality metrics, and trying to determine the *most appropriate* model that fits a given set of data; it is important to bear in mind that *trying to determine the best model that fits a given set of data* is an ill-posed problem. This is because any predictive model is instrumental to a given purpose. Therefore a model that may be suitable to support a given decision, may perform poorly (or simply become irrelevant) if employed in the context of a different decision.

It becomes then clear that three elements are necessary to make sure the problem we are dealing with is well-posed: the data, a predictive model that captures *salient features* of the data, and the associated prescriptive challenge, which is the decision making problem which the predictive model aims to support.

As we have seen in the previous chapters, in inventory control there are essentially three decisions that must be considered in each period: when to review the inventory, when to place an order, and how much to order. To determine answers to each of these questions, one needs, to the very least, forecasts of the future *level* — that is the mean value μ — of the customer demand per period; these can be used in the context of deterministic decision support models such as those we have considered in previous chapters. Ideally, however, to enable probabilistic reasoning one is also interested in obtaining an estimation of the *forecast errors* associated with these estimates — that is the standard deviation σ of the model residuals, which we have shown how to compute in this chapter.

In essence, no matter what forecasting model we adopt to predict customer demand, what we aim to produce are two values: μ_t and σ_t for each period t in our planning horizon. In the next section, we will see how these estimates can then be used to build stochastic decision support models to address specific challenges in the realm of prescriptive inventory analytics.

Finally, it is important to point out that separating a given challenge into predictive and prescriptive subproblems is a convenient *divide and conquer* strategy, which may however lead to suboptimal decisions. Latest research in inventory control [see e.g. Rossi et al., 2014a, Levi et al., 2015] aims at addressing these questions in an integrated manner rather than independently.

Stochastic Inventory Control

https://doi.org/10.11647/OBP.0252.04

Introduction

In this chapter, we discuss inventory control in a stochastic setting. We first introduce the Newsvendor problem, which is the simplest possible stochastic inventory system that can be conceived. Then we survey service level constraints and penalty cost schemes that can be adopted for modelling stochastic inventory systems. Next, we show how to model and simulate a stochastic inventory system running costs. Equipped with these notions, we extend the Newsvendor problem to a multi-period setting. Central to stochastic inventory theory is the notion of *control policy*. A control policy is a rule that establishes when inventory should be reviewed and orders issued, and how large an order should be. The rest of this chapter is devoted to presenting a range of control policies that are commonly adopted in inventory control and that can be used to control a number of well-known inventory systems.

Topics

The Newsvendor

The name of this model derives from its analogy to the problem faced by a newsvendor who purchases newspapers at the beginning of the day before attempting to sell them in the street. Arrow[42] attributes the development of the Newsvendor model to Edgeworth,[43] who used the central limit theorem to determine the optimal cash reserves to satisfy random withdrawals from depositors.

[42] Kenneth J. Arrow. *Studies in the mathematical theory of inventory and production*. Stanford Univ. Pr., 1977.

[43] Francis Y. Edgeworth. The mathematical theory of banking. *Journal of the Royal Statistical Society*, 51(1):113–127, 1888.

The Newsvendor is the simplest stochastic inventory problem one may conceive. It concerns a single item type over a planning horizon that comprises a single period. There is a single opportunity to order, at the beginning of the period, to meet a random demand d that follows a known cumulative distribution function (CDF) F, and that is observed after the order is received (Fig. 89).

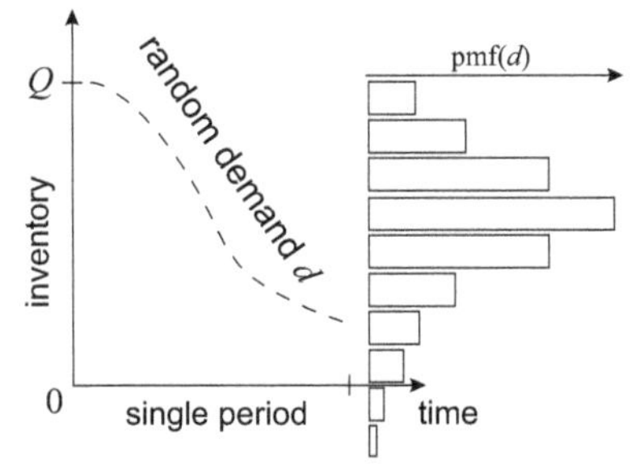

Fig. 89 The Newsvendor problem.

Items can be ordered at a purchasing cost c per unit; they are sold at a selling price p per unit; and, if some items remain unsold at the end of the day, their salvage value is s per unit. These parameters must satisfy $s < c < p$ for the problem to make sense.

Purchasing cost	c	per unit
Selling price	p	per unit
Salvage value	s	per unit

Let Q denote the quantity ordered at the beginning of the period,

- the ordering cost is cQ;
- $\min(Q,d)$ units are sold at p per item; and
- $\max(Q-d,0)$ units are salvaged at value s per item.

The profit function is

$$P(Q) \triangleq p\mathrm{E}[\min(Q,d)] + s\mathrm{E}[\max(Q-s,0)] - cQ$$

where E denotes the expected value operator.

Observe that $\mathrm{E}[\min(Q,d)] = \mathrm{E}[d] - \mathrm{E}[\max(d-Q,0)]$, then

$$\begin{aligned} P(Q) &\triangleq p(\mathrm{E}[d] - \mathrm{E}[\max(d-Q,0)]) + s\mathrm{E}[\max(Q-d,0)] - cQ \\ &= (p-c)\mathrm{E}[d] - p\mathrm{E}[\max(d-Q,0)] + s\mathrm{E}[\max(Q-d,0)] - c\mathrm{E}[Q-d] \\ &= (p-c)\mathrm{E}[d] - (p-c)\mathrm{E}[\max(d-Q,0)] - (c-s)\mathrm{E}[\max(Q-d,0)] \end{aligned}$$

since $\max(x,0) - \max(-x,0) = x$.

Term $(p-c)\mathrm{E}[d]$ is constant, therefore maximising $P(Q)$ is equivalent to minimising

$$C(Q) \triangleq (p-c)\mathrm{E}[d-Q]^+ + (c-s)\mathrm{E}[Q-d]^+ \geq 0,$$

where $[x]^+ \triangleq \max(x,0)$.

In essence, if demand is less than Q, the decision maker faces a so-called "overage" cost $o \triangleq c - s$ per item. If demand exceeds Q, she faces an opportunity cost equal to the missed profit $u \triangleq p - c$ per item, which we shall call "underage" cost.

Overage cost (o)	$c-s$	per unit
Underage cost (u)	$p-c$	per unit

Consider a random variable ω with CDF F, and a scalar x.

Definition 19. $\mathrm{E}[\omega - x]^+$ *is the first order loss function.*[44]

Definition 20. $\mathrm{E}[x - \omega]^+$ *is the complementary first order loss function.*

Lemma 35. $\mathrm{E}[\omega - x]^+ = \mathrm{E}[x - \omega]^+ - (x - \mathrm{E}[\omega])$.

Proof. See [Rossi et al., 2014b, Lemma 3]. □

[44] Roberto Rossi, S. Armagan Tarim, Steven Prestwich, and Brahim Hnich. Piecewise linear lower and upper bounds for the standard normal first order loss function. *Applied Mathematics and Computation*, 231:489–502, 2014b.

As we will see, these two convex [Rossi et al., 2014b, Lemma 4] functions play a key role in stochastic inventory control.

Lemma 36. $C(Q)$ *is convex.*

Proof. The overage cost o and the underage cost u are positive; both $\mathrm{E}[x - d]^+$ and $\mathrm{E}[d - x]^+$ are convex [Rossi et al., 2014b, Lemma 4]; and thus $C(Q)$ is the sum of two convex functions (Fig. 90). □

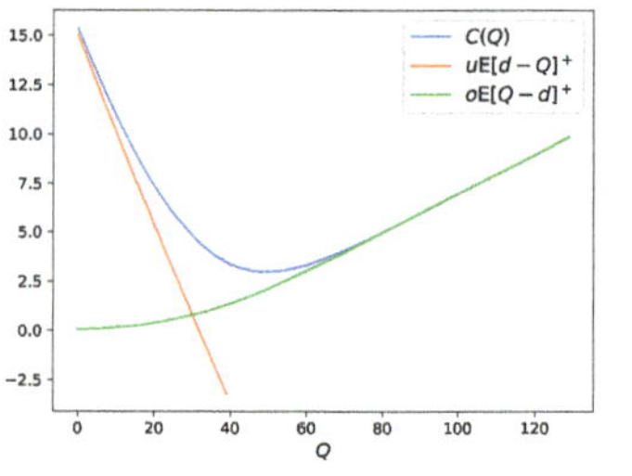

Fig. 90 The Newsvendor problem: the cost function $C(Q)$ and its components $u\mathrm{E}[d - Q]^+$ and $o\mathrm{E}[Q - d]^+$.

Lemma 37. $\mathrm{E}[x - \omega]^+ = \int_{-\infty}^{x} F(t)\mathrm{d}t$.

Proof. See [Rossi et al., 2014b, Lemma 2]. □

Lemma 38 (Critical fractile). *Let* $Q^* \triangleq \min C(Q)$, *then*

$$Q^* = F^{-1}\left(\frac{u}{o+u}\right),$$

where F^{-1} *is the inverse cumulative distribution function of* d. *This is known as the critical fractile solution.*[45]

[45] Kenneth J. Arrow, Theodore Harris, and Jacob Marschak. Optimal inventory policy. *Econometrica*, 19(3):250–272, 1951.

Proof. Rewrite

$$\begin{aligned} C(Q) &= u\mathrm{E}[d-Q]^+ + o\mathrm{E}[Q-d]^+ \\ &= u(\mathrm{E}[Q-\omega]^+ - (Q - \mathrm{E}[\omega])) + o\mathrm{E}[Q-d]^+ \\ &= (u+o)\mathrm{E}[Q-\omega]^+ - u(Q - \mathrm{E}[\omega]). \end{aligned}$$

We take the derivative of $C(Q)$, and by applying Lemma 37, we obtain $C'(Q) = (u+o)F(Q) - u$. Since $C(Q)$ is convex (Lemma 39), to find its minimum we let $C'(Q) = 0$;[46] and by inverting the cumulative distribution function F, we obtain the desired result. □

[46] This is known as the "first order condition."

If the random demand d is discrete, the global minimum can be easily found by analysing the forward differences of $C(Q)$.

We next implement the Newsvendor in Python (Listing 68).

```
import scipy.integrate as integrate
from scipy.stats import norm
from scipy.optimize import minimize

class Newsvendor:
   def __init__(self, instance):
      self.mean, self.std, self.o, self.u = instance["mean"], instance["std"],
         instance["o"], instance["u"]

   def crit_frac_solution(self): # critical fractile
      return norm.ppf(self.u/(self.o+self.u), loc=self.mean, scale=self.std)

   def cfolf(self,Q): # complementary first order loss function
      return integrate.quad(lambda x: norm.cdf(x, loc=self.mean, scale=self.std), 0,
         Q)[0]

   def C(self,Q): # C(Q)
      return (self.o+self.u)*self.cfolf(Q)-self.u*(Q - self.mean)

   def optC(self): # min C(Q)
      return minimize(self.C, 0, method='Nelder-Mead')
```

```
instance = {"o" : 1, "u": 5, "mean" :
   10, "std" : 2}
nb = Newsvendor(instance)
print("Q^*=" +
   str(nb.crit_frac_solution()))
print("C(Q^*)=" +
   str(nb.C(nb.crit_frac_solution())))
print(nb.optC())
```

Listing 68 A Newsvendor instance.

Service level constraints in inventory systems

In this section we survey the different types of service level constraints that are typically adopted in inventory control. While doing so, we illustrate applications to the Newsvendor problem.

α SERVICE LEVEL (NO STOCKOUT PROBABILITY): this is defined as the "probability of no stockout per order cycle." In the case of the Newsvendor problem, recall that the critical fractile solution is defined as

$$Q^* = F^{-1}\left(\frac{u}{o+u}\right).$$

But F^{-1} is the cumulative distribution of the demand, therefore if we let $\alpha \triangleq u/(o+u)$, Q^* becomes the order quantity that guarantees a probability of no stockout per order cycle equal to α. Every choice of u and o is therefore equivalent to a given α service level. Order quantities ensuring an arbitrary α service level, can be obtained by inverting the cumulative distribution of the demand (Fig. 91).

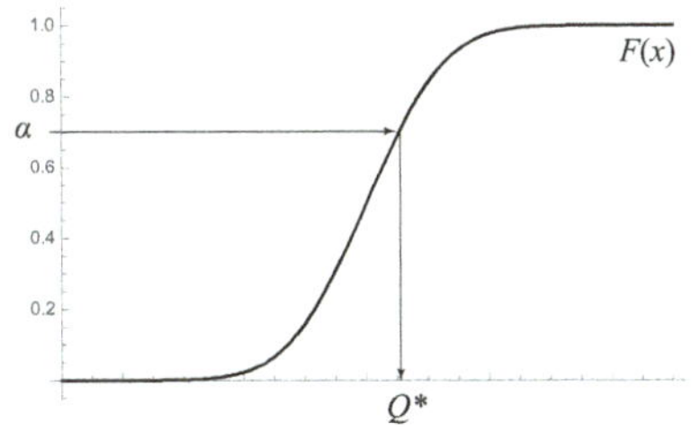

Fig. 91 Inverting the cumulative distribution $F(x)$ of the demand to determine the order quantity Q^* that ensures a given α service level.

β SERVICE LEVEL (FILL RATE): this is defined as the "expected fraction of demand that can be satisfied immediately from stock on hand." The order quantity Q^* that ensures a given β service level can be easily computed by leveraging the complementary first order loss function

$$\frac{\mathrm{E}[Q^* - \omega]^+}{\mathrm{E}[\omega]} = \beta.$$

In Listing 69 we extend the Newsvendor class with α and β service levels, which are the most commonly used in practice.

```
class Newsvendor:
    ...
    def critical_fractile(self):
        return self.u/(self.o+self.u)

    def alpha(self, Q):
        return norm.cdf(Q,
                loc=self.mean,
                scale=self.std)

    def beta(self, Q):
        return self.cfolf(Q)/self.mean
```

Listing 69 Extending the Newsvendor class with service levels.

γ SERVICE LEVEL (READY RATE): this is defined as the "expected fraction of time with positive stock on hand." We shall illustrate this service level for a Newsvendor problem subject to Poisson demand. As discussed in our Appendix on Poisson processes, the inter-arrival times between two Poisson arrivals originating from a Poisson process with mean λ follow an exponential distribution with mean λ^{-1} (and thus rate parameter λ). The sum of n exponentially distributed random variables with rate parameter λ follows an Erlang distribution with shape parameter n and rate parameter λ, whose mean is n/λ. Let $Q^* = n$, the ready rate can be computed as $\gamma = Q^*/\lambda$ — essentially the sum of n inter-arrival times whose expected value is the sought ready rate $0 \leq \gamma \leq 1$. Intuitively, λ is the expected number of poisson arrivals per period, hence Q/λ is the expected fraction of a period required to observe Q arrivals.

SHORTAGE COSTS. Finally, recall that in general, there are three possible strategies for charging stockout/backorder penalty cost in inventory systems: we may charge a fixed or per unit cost once, when one or more units of demand are not met; or we may charge this cost per unit of demand short per time period, until an order arrives and such unit of demand is served.

Simulating stochastic inventory systems

In this section, we illustrate how to modify our discrete simulation framework (p. 39) to accommodate a stochastic demand.

First, it is necessary to slightly modify the DES engine as follows.

```
class EventWrapper():
   def __init__(self, event):
      self.event = event
   def __lt__(self, other):
      return self.event.priority < other.event.priority

class DES():
   def __init__(self, end):
      self.events, self.end, self.time = PriorityQueue() , end, 0

   def start(self):
      while True:
         event = self.events.get()
         self.time = event[0]
         if self.time <= self.end: # this is now <=
            event[1].event.end()
         else:
            break

   def schedule(self, event: EventWrapper, time_lag: int):
      self.events.put((self.time + time_lag, event))
```

We also implement penalty cost accounting in class Warehouse;

```
class Warehouse:
   def __init__(self, inventory_level, fixed_ordering_cost, holding_cost,
        penalty_cost):
      self.i, self.K, self.h, self.p = inventory_level, fixed_ordering_cost,
           holding_cost, penalty_cost
      self.o = 0 # outstanding_orders
      self.period_costs = defaultdict(int) # a dictionary recording period costs

   def receive_order(self, Q, time):
      self.review_inventory(time)
      self.i, self.o = self.i + Q, self.o - Q
      self.review_inventory(time)

   def order(self, Q, time):
      self.review_inventory(time)
      self.period_costs[time] += self.K # incur ordering cost and store it
      self.o += Q
      self.review_inventory(time)

   def on_hand_inventory(self):
      return max(0,self.i)

   def backorders(self):
      return max(0,-self.i)

   def issue(self, demand, time):
      self.review_inventory(time)
      self.i = self.i-demand

   def inventory_position(self):
      return self.o+self.i

   def review_inventory(self, time):
      try:
         self.levels.append([time, self.i])
         self.on_hand.append([time, self.on_hand_inventory()])
         self.positions.append([time, self.inventory_position()])
      except AttributeError:
         self.levels, self.on_hand = [[0, self.i]], [[0, self.on_hand_inventory()]]
         self.positions = [[0, self.inventory_position()]]

   def incur_end_of_period_costs(self, time): # incur holding and penalty costs
      self._incur_holding_cost(time)
      self._incur_penalty_cost(time)

   def _incur_holding_cost(self, time): # incur holding cost and store it
      self.period_costs[time] += self.on_hand_inventory()*self.h

   def _incur_penalty_cost(self, time): # incur penalty cost and store it
      self.period_costs[time] += self.backorders()*self.p
```

and we modify the `EndOfPeriod` event accordingly

```
class EndOfPeriod:
    def __init__(self, des: DES, warehouse: Warehouse):
        self.w = warehouse # the warehouse
        self.des = des # the Discrete Event Simulation engine
        self.priority = 0 # denotes a high priority

    def end(self):
        self.w.incur_end_of_period_costs(self.des.time)
        self.des.schedule(EventWrapper(EndOfPeriod(self.des, self.w)), 1)
```

To model a Poisson demand with rate λ units per period rather than a deterministic demand of λ units per period, in line with what we discuss in our Appendices on Poisson processes and on Discrete Event Simulation (p. 170), we slightly modify method `end` in class `CustomerDemand` as follows.

```
class CustomerDemand:
    def __init__(self, des: DES, demand_rate: float, warehouse: Warehouse):
        self.d = demand_rate # the demand rate per period
        self.w = warehouse # the warehouse
        self.des = des # the Discrete Event Simulation engine
        self.priority = 2 # denotes a low priority

    def end(self):
        self.w.issue(1, self.des.time) # issue one unit of demand
        self.des.schedule(EventWrapper(self), np.random.exponential(1/self.d)) #
            schedule another demand with an exponentially distributed delay
```

Events Order and ReceiveOrder are now assigned a medium priority (`self.priority = 1`).

We consider the same system simulated in Example 2.

Example 29. *We simulate operations of a simple inventory system by leveraging the Python code in Listings 70 and 71. The warehouse initial inventory is 10 units. The customer demand follows a Poisson distribution with a rate of 10 unit per period. We simulate $N = 20$ periods. We order 50 units in periods 1, 5, 10, and 15; the delivery lead time is 1 period. The fixed ordering cost is 100, the per unit inventory holding cost is 1.*

```
def plot_inventory(values, label):
    # data
    df=pd.DataFrame({'x':
            np.array(values)[:,0], 'fx':
            np.array(values)[:,1]})

    # plot
    plt.xticks(range(len(values)),
            range(1,len(values)+1))
    plt.xlabel("t")
    plt.ylabel("items")
    plt.plot( 'x', 'fx', data=df,
            linestyle='-', marker='',
            label=label)
```

Listing 70 The `plot_inventory` function.

```
np.random.seed(1234) # use common random numbers to ensure replicability
instance = {"inventory_level": 10, "fixed_ordering_cost": 100, "holding_cost": 1,
        "penalty_cost": 5}
w = Warehouse(**instance)

N = 20 # planning horizon length
des = DES(N)

d = CustomerDemand(des, 10, w)
des.schedule(EventWrapper(d), 0) # schedule a demand immediately

lead_time = 1
o = Order(des, 50, w, lead_time)
for t in range(0,20,5):
    des.schedule(EventWrapper(o), t) # schedule orders
des.schedule(EventWrapper(EndOfPeriod(des, w)), 1) # schedule EndOfPeriod at the
        end of the first period
des.start()

print("Period costs: "+str([w.period_costs[e] for e in w.period_costs]))
print("Average cost per period: "+ '%.2f' % (sum([w.period_costs[e] for e in
        w.period_costs])/len(w.period_costs)))

plot_inventory(w.positions, "inventory position")
plot_inventory(w.levels, "inventory level")
plt.legend(loc="lower right")
plt.show()
```

Listing 71 Simulating the behaviour of a warehouse in Python. To ensure replicability, we leverage common random numbers [Kahn and Marshall, 1953].

After simulating the system, we find that the average cost per unit time is 51, this is higher than the cost (40) observed when demand is deterministic and equal to λ per period. The behaviour of the inventory system in terms of inventory level and inventory position at the end of each period is shown in Fig. 92.

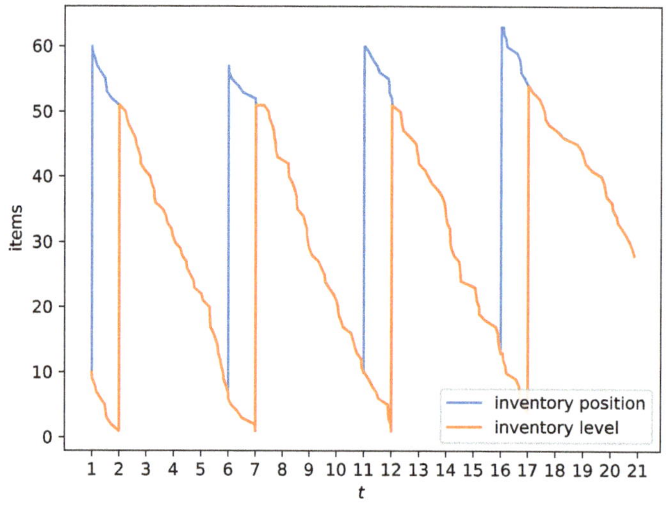

Fig. 92 Simulating the behaviour of a warehouse in Python subject to stochastic demand: inventory level and inventory position at the end of each period $t \in \{1, 20\}$ when the initial inventory level is 10. Demand now follows a Poisson distribution with a rate of 10 units per period.

We next use the discrete simulation code just presented to simulate a Newsvendor problem with $o = 1$, $u = 5$, and Poisson demand with $\lambda = 100$.

```
def simulate_newsvendor():
    instance = {"inventory_level": 0, "fixed_ordering_cost": 0, "holding_cost": 1,
         "penalty_cost": 5}
    w = Warehouse(**instance)

    N = 1 # planning horizon length
    des = DES(N)

    d = CustomerDemand(des, 100, w)
    des.schedule(EventWrapper(d), 0) # schedule a demand immediately

    lead_time = 0
    o = Order(des, 110, w, lead_time)
    des.schedule(EventWrapper(o), 0) # schedule a single order at time 0
    des.schedule(EventWrapper(EndOfPeriod(des, w)), 1) # schedule EndOfPeriod with a
         delay of one period

    des.start()

    print("Period costs: "+str([w.period_costs[e] for e in w.period_costs]))
    print("Total cost: "+ '%.2f' % (sum([w.period_costs[e] for e in w.period_costs])))

    return sum([w.period_costs[e] for e in w.period_costs])

np.random.seed(1234)
replications = 1000
print("Simulated cost: " +str(sum([simulate_newsvendor() for k in
     range(replications)])/replications))
```

By noting that a Poisson demand with $\lambda = 100$ is roughly equivalent to a normally distributed demand with $\mu = 100$ and $\sigma = 10$, by using analytical results previously presented for the Newsvendor model, we find that $Q^* = 110$ and $C(Q^*) = 14.99$. The simulated cost is 15.01, which closely approximates the analytical solution.

The multi-period Newsvendor

This is a variant of the Newsvendor model in which the planning horizon comprises T periods (Fig. 93). In each period $t = 1, \dots, T$, we observe a random demand d_t with known probability distribution. There is still a single opportunity to order, so that an order quantity Q can be ordered only at the beginning of the planning horizon. However, inventory overage ($o \triangleq c - s$) and underage ($u \triangleq p - c$) costs are now incurred at the end of each period.

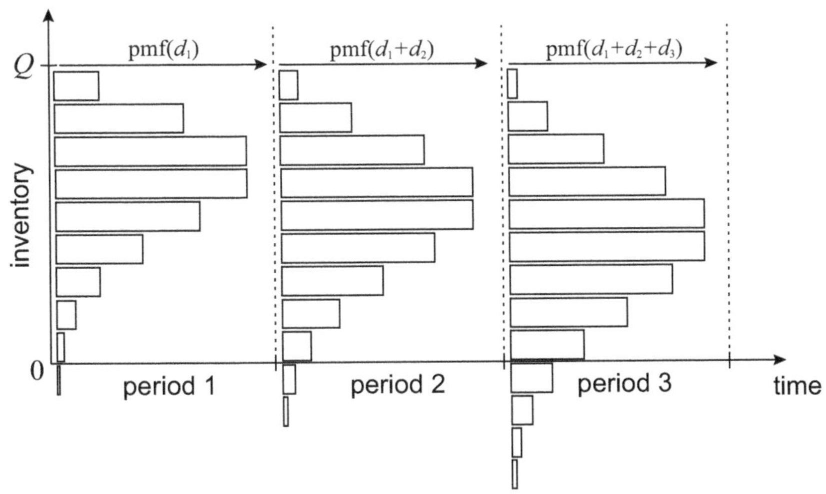

Fig. 93 The multi-period Newsvendor problem.

Let Q denote the quantity ordered at the beginning of the period, our goal is to minimise

$$C(Q) \triangleq \sum_{t=1}^{T} (p-c)\mathrm{E}[d_{1..t} - Q]^+ + (c-s)\mathrm{E}[Q - d_{1..t}]^+ \geq 0,$$

where $[x]^+ \triangleq \max(x, 0)$; and $d_{1..t} \triangleq d_1 + d_2 + \ldots + d_t$.

Lemma 39. *$C(Q)$ is convex.*

Proof. For every t, function $(p-c)\mathrm{E}[d_{1..t} - Q]^+ + (c-s)\mathrm{E}[Q - d_{1..t}]^+$ is equivalent to a traditional Newsvendor cost function; therefore $C(Q)$ is the sum of T convex functions. □

Lemma 40 (Critical fractile). *Let $Q^* \triangleq \min C(Q)$, then*

$$\sum_{t=1}^{T} F_{1..t}(Q^*) = \frac{Tu}{o+u}, \qquad (29)$$

where $F_{1..t}$ is the cumulative distribution function of $d_1 + d_2 + \ldots + d_t$. The optimal order quantity can be found by finding Q^* that solves Eq. 29.

Proof. A proof of this result is provided in [Askin, 1981, p. 133]. □

Once more, if the random demand d is discrete, the global minimum can be found by analysing the forward differences of $C(Q)$.

The multi-period Newsvendor can be implemented in Python as shown below.

```
from itertools import accumulate
from scipy.stats import poisson
import scipy.integrate as integrate
from scipy.optimize import minimize

class MultiPeriodNewsvendor:
   def __init__(self, instance):
      self.mean, self.o, self.u = instance["mean"], instance["o"], instance["u"]

   def cfolf(self, Q, d): # complementary first order loss function
      return integrate.quad(lambda x: poisson.cdf(x, d), 0, Q)[0]

   def folf(self,Q): # first order loss function
      return self.cfolf(Q)-self.u*(Q - self.mean)

   def C(self, Q): # C(Q)
      return sum([(self.o+self.u)*self.cfolf(Q, d)-self.u*(Q - d) for d in
            accumulate(self.mean)])

   def optC(self): # min C(Q)
      return minimize(self.C, 0, method='Nelder-Mead')

   def verify_fractile_solution(self, Q):
      T = len(self.mean)
      critical_fractile = T*self.u/(self.u+self.o)
      return sum([poisson.cdf(Q, d) for d in accumulate(self.mean)]) -
            critical_fractile < 0.1

instance = {"o" : 1, "u": 5, "mean" : [10,10,10]}
nb = MultiPeriodNewsvendor(instance)
res = nb.optC()
print(res)
print("Verify critical fractile: "+str(nb.verify_fractile_solution(res.x[0])))
```

Example 30. *We consider an instance comprising $T = 3$ periods. Demand in each period t follows a Poisson distribution with $\lambda_t = 10$. After solving this instance, the optimal order quantity is $Q^* = 30$ and $C(Q^*) = 43.30$. Q^* has been obtained by using Nelder-Mead optimisation algorithm [Nelder and Mead, 1965]. However, the cost also verifies via function* `verify_fractile_solution` *that Q^* is a solution to Eq. 29.*

We can employ once more DES to simulate the cost of this inventory system. The Python code is shown below.

```
def simulate_newsvendor():
   instance = {"inventory_level": 0, "fixed_ordering_cost": 0, "holding_cost": 1,
         "penalty_cost": 5}
   w = Warehouse(**instance)

   N = 3 # planning horizon length
   des = DES(N)

   d = CustomerDemand(des, 10, w)
   des.schedule(EventWrapper(d), 0) # schedule a demand immediately

   lead_time = 0
   o = Order(des, 30, w, lead_time)
   des.schedule(EventWrapper(o), 0) # schedule a single order at time 0
   des.schedule(EventWrapper(EndOfPeriod(des, w)), 1) # schedule EndOfPeriod with a
         delay of one period
   des.start()

   print("Period costs: "+str([w.period_costs[e] for e in w.period_costs]))
   print("Total cost: "+ '%.2f' % (sum([w.period_costs[e] for e in w.period_costs])))

   return sum([w.period_costs[e] for e in w.period_costs])

np.random.seed(1234)
replications = 10000
print("Simulated cost: " +str(sum([simulate_newsvendor() for k in
      range(replications)])/replications))
```

The simulated value of $C(Q^*)$ is 43.80, which is close to the result previously obtained by using Nelder-Mead optimisation algorithm.

The base-stock policy

We now consider the setting analysed in the multi-period Newsvendor problem, but we relax the single-order assumption. The aim is then to control inventory of a single item type over a planning horizon that comprises T periods. As in the multi-period Newsvendor problem, we operate in a *periodic-review* setting. This means that there is an opportunity to order at the beginning of each period t, to meet a random demand d_t that follows a known probability distribution, which may differ from period to period; and that inventory overage (also known as holding h) cost and inventory underage (also known as backorder penalty cost p) are incurred at the end of each period. Unmet demand at the end of each period is carried over (*backordered*) to the next period, and met as soon as the next replenishment arrives.

Since at the beginning of each period t, we have an opportunity to replenish our stock, and since we incur no fixed cost for this, the problem is essentially equivalent to solving a set of T independent single-period Newsvendor problems.

Let I_{t-1} be the closing inventory level at the end of period $t-1$, by leveraging the critical fractile solution presented in Lemma 38, we define the order-up-to-level for period t as follows

$$S_t \triangleq F_t^{-1}\left(\frac{p}{h+p}\right),$$

where F_t^{-1} is the inverse cumulative distribution function of d_t. The optimal action at the beginning of period t is therefore to order $Q_t^* = \max\{S_t - I_{t-1}, 0\}$. In essence, at the beginning of period t, we *order-up-to* S_t. This control policy is known as the *base-stock policy*.

Example 31. *Let us consider a planning horizon comprising $T = 5$ periods. Demand d_t in each period $t = 1, \ldots, T$ follows a Poisson distribution with mean $\lambda_t = 10$; for the sake of simplicity we here assume, without loss of generality, that all λ_t are equal. Inventory holding cost is $h = 1$, and inventory backorder penalty cost is $p = 5$. We solve five separate Newsvendor problems, one for each period, we obtain $S_t = 13$, for all t. The behaviour of the system when initial inventory is 10 is shown in Fig. 94*

Simulating a base-stock policy only requires some minor adjustments to our DES Python code. In particular, we shall replace the `Order` class with the following `OrderUpTo` class.

```
class OrderUpTo:
    def __init__(self, des: DES, S: float, warehouse: Warehouse, lead_time: float):
        self.S = S # the order-up-to-level
        self.w = warehouse # the warehouse
        self.des = des # the Discrete Event Simulation engine
        self.lead_time = lead_time
        self.priority = 1 # denotes a medium priority

    def end(self):
        Q = self.S - self.w.inventory_position()
        self.w.order(Q, self.des.time)
        self.des.schedule(EventWrapper(ReceiveOrder(self.des, Q, self.w)),
             self.lead_time)
```

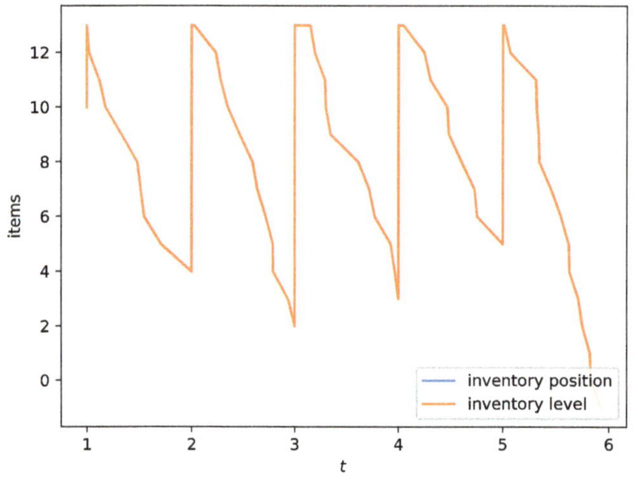

Fig. 94 Simulating the base-stock policy in Example 31.

The problem can then be simulated as follows.

```
np.random.seed(1234)
instance = {"inventory_level": 10, "fixed_ordering_cost": 0, "holding_cost": 1,
     "penalty_cost": 5}
w = Warehouse(**instance)

N = 5 # planning horizon length
des = DES(N)

d = CustomerDemand(des, 10, w)
des.schedule(EventWrapper(d), 0) # schedule a demand immediately

lead_time = 0
o = OrderUpTo(des, 13, w, lead_time)
for t in range(5):
   des.schedule(EventWrapper(o), t) # schedule orders
des.schedule(EventWrapper(EndOfPeriod(des, w)), 1) # schedule EndOfPeriod at the
     end of the first period
des.start()

print("Period costs: "+str([w.period_costs[e] for e in w.period_costs]))
print("Average cost per period: "+ '%.2f' % (sum([w.period_costs[e] for e in
     w.period_costs])/len(w.period_costs)))

plot_inventory(w.positions, "inventory position")
plot_inventory(w.levels, "inventory level")
plt.legend(loc="lower right")
plt.show()
```

Lost sales. If we assume that unmet demand is *lost* at the end of each period, as opposed to being backordered, the computation of S_t does not change. However, the optimal action at the beginning of period t is now to order $Q_t^* = \max\{S_t - \max\{I_{t-1}, 0\}, 0\}$.

Positive order lead time. If an order is delivered after a positive lead time of l periods, the problem can be solved by leveraging the solution to a multi-period Newsvendor over periods $t, \ldots, t+l$ while computing the order-up-to-levels. In particular,

$$S_t \triangleq \left\{ S \,\middle|\, \sum_{t=1}^{T} F_{t..t+l}(S) = \frac{Tp}{h+p} \right\},$$

where $F_{t..t+l}$ is the inverse cumulative distribution function of $d_t + \ldots, d_{t+l}$; this accounts for the demand variability over lead time.

To compute the associated optimal order quantity, first we need to establish the order of events. We shall assume that, at the beginning of a period, outstanding orders that are due to be delivered in such period are received; then inventory is reviewed; and new orders are issued. Let O_t denote the total outstanding order quantity at the beginning of period t, when inventory is reviewed: these are all the orders already issued, but not yet received. Define the *inventory position* $P_t \triangleq O_t + I_{t-1}$, where I_{t-1} is the inventory level at the end of period t. Then the optimal action at the beginning of period t is to order $Q_t^* = \max\{S_t - P_t, 0\}$.

Example 32. *We consider the instance in Example 31 and set the order delivery lead time to 1 period. By solving a multi-period Newsvendor over a planning horizon of $T = 2$ periods with Poisson demand $\lambda_t = 10$ in each of these, we obtain $S_t = 20$. The system behaviour is simulated in Fig. 95*

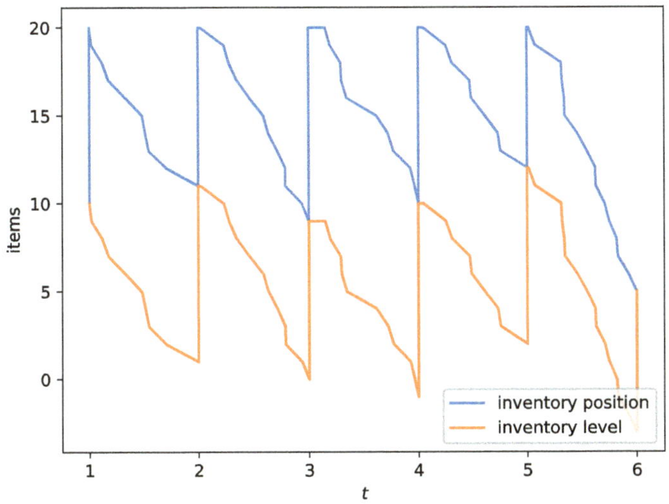

Fig. 95 Simulating the base-stock policy in Example 31 under an order delivery lead time $l = 1$. If we simulate a large number of periods (e.g. 500), we can observe that the average cost per period is now higher (11.14) than that observed for the zero lead time case (4.97). A positive lead time makes it more expensive to control the system.

```
np.random.seed(1234)
instance = {"inventory_level": 10, "fixed_ordering_cost": 0, "holding_cost": 1,
      "penalty_cost": 5}
w = Warehouse(**instance)

N = 5 # planning horizon length
des = DES(N)
d = CustomerDemand(des, 10, w)
des.schedule(EventWrapper(d), 0) # schedule a demand immediately

lead_time = 1
o = OrderUpTo(des, 20, w, lead_time)
for t in range(N):
    des.schedule(EventWrapper(o), t) # schedule orders
des.schedule(EventWrapper(EndOfPeriod(des, w)), 1) # schedule EndOfPeriod at the
      end of the first period
des.start()

print("Period costs: "+str([w.period_costs[e] for e in w.period_costs]))
print("Average cost per period: "+ '%.2f' % (sum([w.period_costs[e] for e in
      w.period_costs])/len(w.period_costs)))

plot_inventory(w.positions, "inventory position")
plot_inventory(w.levels, "inventory level")
plt.legend(loc="lower right")
plt.show()
```

The modified base-stock policy

We now consider once more the setting analysed in the multi-period Newsvendor problem, but in addition to relaxing the single-order assumption, we assume that an order capacity constraint is in place, so that at the beginning of each period, it is only possible to order a quantity less or equal to B.

It is possible to show that the optimal policy to this periodic review problem is a so-called *modified base-stock policy* in which, in each period, we order up to the order-up-to-level S_t, or as close as possible to it, given the ordering capacity.[47] Therefore

$$Q_t^* = \max\{\min\{S_t - I_{t-1}, B\}, 0\}.$$

This result directly follows from the following lemma.

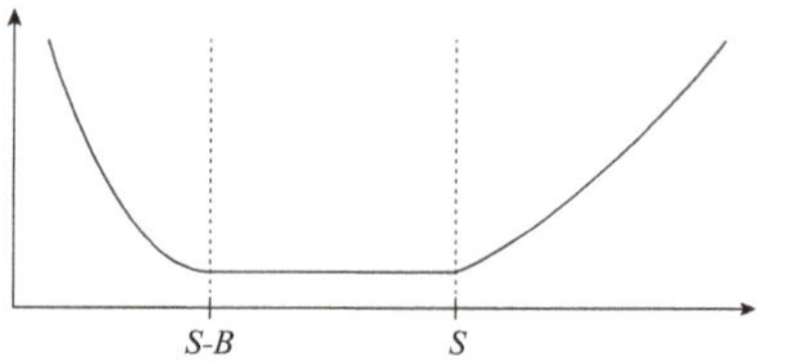

Fig. 96 Plot of $\min_{y\in[x,x+B]} f(y)$.

[47] Awi Federgruen and Paul Zipkin. An inventory model with limited production capacity and uncertain demands I. The average-cost criterion. *Mathematics of Operations Research*, 11 (2):193–207, 1986.

Lemma 41. *Let f be convex, and S be a minimizer of f, then*

$$\min_{y\in[x,x+B]} f(y) = \begin{cases} f(x) & S \leq x \\ f(S) & S - B \leq x \leq S \\ f(x+B) & x \leq S - B \end{cases}$$

Proof. A proof (Fig. 96) is discussed in [Karush, 1959]. □

Example 33. *We consider the instance in Example 31 and an ordering capacity $B = 13$. The order-up-to-levels are not affected by the ordering capacity, and they remain equal to $S_t = 13$, for all t.*

The system can be once more simulated by making a few tweaks to our DES code. In particular, we replace the `OrderUpTo` class with the following `ModifiedOrderUpTo` class in Listing 72. Finally, we amend the rest of the code as shown below.

```
np.random.seed(1234)
instance = {"inventory_level": 10, "fixed_ordering_cost": 0, "holding_cost": 1,
     "penalty_cost": 5}
w = Warehouse(**instance)

N = 5 # planning horizon length
des = DES(N)
d = CustomerDemand(des, 10, w)
des.schedule(EventWrapper(d), 0) # schedule a demand immediately

lead_time, capacity = 0, 13
o = ModifiedOrderUpTo(des, 13, w, lead_time, capacity)
for t in range(N):
    des.schedule(EventWrapper(o), t) # schedule orders
des.schedule(EventWrapper(EndOfPeriod(des, w)), 1) # schedule EndOfPeriod at the
     end of the first period
des.start()

print("Period costs: "+str([w.period_costs[e] for e in w.period_costs]))
print("Average cost per period: "+ '%.2f' % (sum([w.period_costs[e] for e in
     w.period_costs])/len(w.period_costs)))

plot_inventory(w.positions, "inventory position")
plot_inventory(w.levels, "inventory level")
plt.legend(loc="lower right")
plt.show()
```

```
class ModifiedOrderUpTo:
    def __init__(self, des: DES, S:
            float, warehouse: Warehouse,
            lead_time: float, B: float):
        self.S = S # the
            order-up-to-level
        self.w = warehouse # the
            warehouse
        self.des = des # the Discrete
            Event Simulation engine
        self.lead_time = lead_time
        self.B = B
        self.priority = 1 # denotes a
            medium priority

    def end(self):
        Q = min(self.S -
            self.w.inventory_position(),
            self.B)
        self.w.order(Q, self.des.time)
        self.des.schedule( EventWrapper(
            ReceiveOrder( self.des, Q,
            self.w)), self.lead_time)
```

Listing 72 The `ModifiedOrderUpTo` class.

If we simulate a large number of periods (e.g. 500), we can observe that the average cost per period is now higher (6.25) than that observed for the case in which the capacity costraint is absent (4.97). An order capacity constraint makes it more expensive to control the system.

The (s, S) policy

We consider a single-item single-stocking location stochastic inventory system subject to random demand. The planning horizon is finite and comprises n periods — for convenience periods are labelled in reverse order $n, n-1, \ldots, 1$. The probability density function of the demand d_t in period t is f_t and its cumulative distribution function is F_t. The system operates under the following cost structure: there is a purchasing or ordering cost $c(x)$ for purchasing x units of inventory; a holding cost h that is charged for transferring one unit of inventory from one period to the next; and a shortage cost p that charged for each unit short at the end of a period. Unmet demand at the end of a period is backordered and met as soon as the next replenishment arrives. We shall assume that deliveries occurs only at the beginning of a period and that they are instantaneous.

Let $C_n(x)$ denote the expected total cost attained by an optimal provisioning policy over periods $n, \ldots, 1$ when the inventory at the beginning of period n, before any order is placed, is x; then

$$C_n(x) \triangleq \min_{x \leq y} \left\{ c(y-x) + L_n(y) + \int_0^\infty C_{n-1}(y-\omega) f_n(\omega) \mathrm{d}\omega \right\};$$

where $C_0 \triangleq 0$, and

$$L_n(y) \triangleq \int_0^y h(y-\omega) f_n(\omega) \mathrm{d}\omega + \int_y^\infty p(\omega - y) f_n(\omega) \mathrm{d}\omega.$$

In what follows we shall concentrate on the case in which the ordering cost takes the following structure

$$c(x) \triangleq \begin{cases} 0 & x = 0, \\ K + vx & x > 0. \end{cases}$$

In other words, we now have a fixed ordering cost component K, which is incurred every time an order is placed, regardless of the order size; and a variable ordering cost component v, which is proportional to the size of the order.

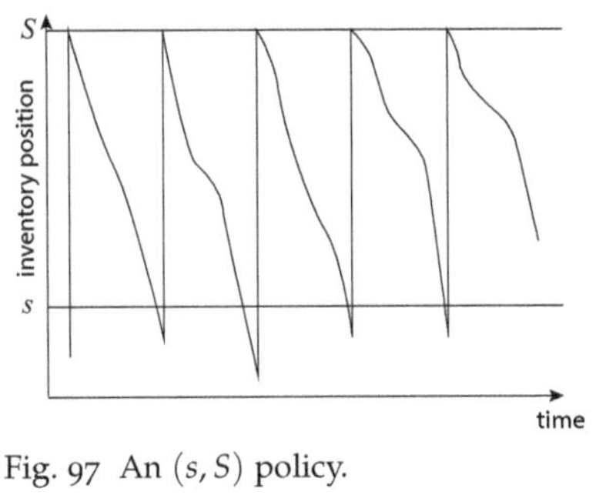

Fig. 97 An (s, S) policy.

The optimality of (s, S) policies

Scarf proved that, if $L_n(y)$ is convex and ordering cost follows the structure of $c(x)$, the optimal policy is defined by a pair of critical numbers (s, S) as follows: at the beginning of each period, if $x < s$ order $S - x$, otherwise, do not order (Fig 97). This control rule is known as the (s, S) policy.[48]

[48] Herbert E. Scarf. Optimality of (s, S) policies in the dynamic inventory problem. In K. J. Arrow, S. Karlin, and P. Suppes, editors, *Mathematical Methods in the Social Sciences*, pages 196–202. Stanford Univ. Pr., 1960.

We first introduce the following definition

Definition 21 (*K*-convexity). *Let $K \geq 0$, $g(x)$ is K-convex if for all x, $a > 0$, and $b > 0$,*

$$\frac{K + g(x+a) - g(x)}{a} \geq \frac{g(x) - g(x-b)}{b}. \tag{30}$$

Scarf's result is based on a study of function

$$G_n(y) \triangleq vy + L_n(y) + \int_0^\infty C_{n-1}(y-\omega)f_n(\omega)\mathrm{d}\omega.$$

Scarf proved that, if $L_n(y)$ is convex, $G_n(y)$ is K-convex and thus satisfy the property in Definition 21.

For illustrative purposes (Fig. 98), it is worth rearranging terms in Eq. 30 as follows

$$K + g(x+a) \geq g(x) + a\frac{g(x)-g(x-b)}{b}.$$

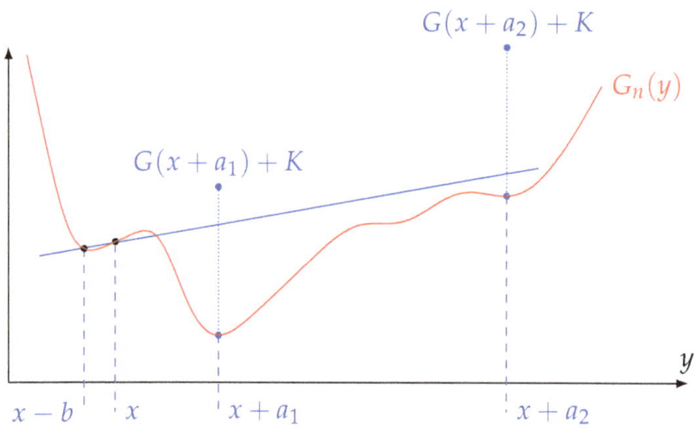

Fig. 98 K-convexity of G_n: let $a, b > 0$, pick two points $(x-b, G(x-b))$ and $(x, G(x))$, draw a straight line passing through them; then for any $x+a$, point $(x+a, G(x+a)+K)$ lies above the straight line.

Without loss of generality, assume now G_n differentiable, and let $b \to 0$; thus obtaining

$$K + G_n(x+a) \geq G_n(x) + aG'_n(x).$$

Consider all points x where $G'_n(x) = 0$, these include all local maxima of G_n. Let S be the global minimiser of G_n, it immediately follows that $K + G_n(S)$ is greater than the value of G_n at any local maximum $x < S$ (Fig. 99).

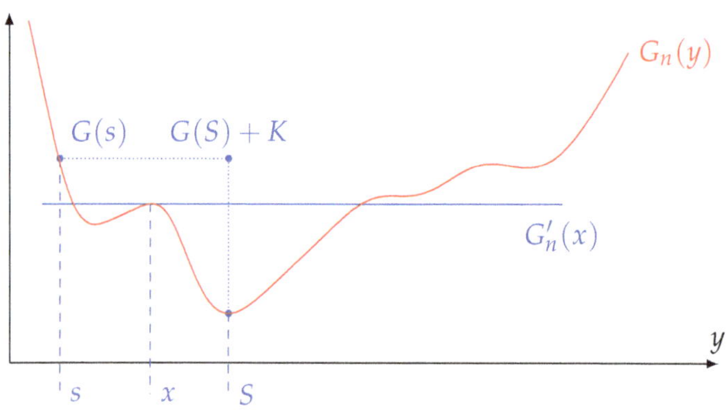

Fig. 99 K-convexity of G_n: $K + G_n(S)$ is greater than the value of G_n at any local maximum $x < S$, thus there exists a unique value s such that $K + G_n(S) = G_n(s)$.

This means that, despite the fluctuations of G_n, there exists a unique value s such that $K + G_n(S) = G_n(s)$, where $S \triangleq \min_y G_n(y)$. In turn, this implies that an (s, S) policy is optimal.

This result, of course, hinges on K-convexity of G_n and C_n, which must be proved.

We first introduce three well-known properties (Lemma 42, Lemma 43, and Lemma 44) of K-convex functions.

Lemma 42. *0-convexity is equivalent to ordinary convexity.*

Lemma 43. *If f and g are K-convex and M-convex, respectively, then $\alpha f + \beta g$ is $(\alpha K + \beta M)$-convex for α and β positive.*

Lemma 44. *If g is K-convex, then $g(x+h)$ is K-convex for all h.*

Scarf's proof proceeds by induction. $C_0 \triangleq 0$; $L(y)$ is convex and hence K-convex (Lemma 42); vy is linear and hence K-convex; $G_1(y)$ is convex (Lemma 43) and hence K-convex (Lemma 42). Assume that $G_2(y), \ldots, G_n(y)$ are K-convex; to show that $G_{n+1}(y)$ is K-convex, it is sufficient to show that $\int_0^\infty C_n(y-\omega) f_n(\omega) \mathrm{d}\omega$ is K-convex; and by applying Lemma 43 and Lemma 44, it is sufficient to show that $C_n(x)$ is K-convex.

Theorem 1. *$C_n(x)$ is K-convex.*

Proof. Observe that under an (s, S) policy, $C_n(x)$ can be expressed as follows

$$C_n(x) \triangleq \begin{cases} K - vx + G_n(S) & x \leq s \\ -vx + G_n(x) & x > s. \end{cases} \tag{31}$$

To prove that $C_n(x)$ is K-convex, we analyse three possible cases (Fig. 100) covering intervals within which x and $x+a$ may lie.

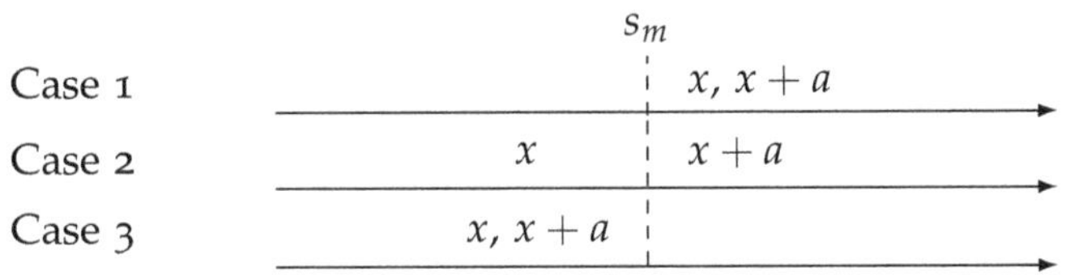

Fig. 100 Cases considered in the proof of Theorem 1.

Case 1: $x > s$. In this region, $C_n(x)$ is equal to a linear function plus a K-convex function, hence it is K-convex.

Case 2: $x < s < x + a$. By applying Definition 21, $C_n(x)$ is K-convex, since $K + C_n(x+a) - C_n(x) - aC_n'(x) \geq 0$, that is

$$K + (-v(x+a) + G_n(x+a)) - (K - vx + G_n(S)) - a(-v + G_n'(S)) \geq 0$$

which is true, since $G_n(x+a) - G_n(S) \geq 0$, for all x and $a > 0$, and $G_n'(S) = 0$, because S is the global minimiser of G_n.

Case 3: $x < x + a < s$. In this region, $C_n(x)$ is linear. □

Observe that this proof applies both to stationary and nonstationary demand. The proof can be also extended to the case of a positive deterministic lead time; this extension is discussed in [Scarf, 1960].

Computing stationary (s, S) policy parameters

We consider a single-item single-stocking location stochastic inventory system subject to stationary stochastic demand over an infinite planning horizon and periodic review. The probability density function of the demand d in any given period is f and its cumulative distribution function is F. The system operates under the following cost structure: there is a purchasing or ordering cost $c(x)$ for purchasing x units of inventory; a holding cost h that is charged for transferring one unit of inventory from one period to the next; and a shortage cost p that charged for each unit short at the end of a period. Unmet demand at the end of a period is backordered and met as soon as the next replenishment arrives. We shall assume that deliveries occurs only at the beginning of a period and that they are instantaneous. The objective is to minimise the expected long run average cost per period.

According to what was discussed in the previous section, an (s, S) policy is optimal for this system. We shall now illustrate how to compute optimal (s, S) policy parameters by following the approach discussed by Zheng and Federgruen.[49]

[49] Yu-Sheng Zheng and Awi Federgruen. Finding optimal (s, S) policies is about as simple as evaluating a single policy. *Operations Research*, 39(4): 654–665, 1991.

At the beginning of any given period, if we have y units in stock, the expected total holding and backordering costs can be expressed as

$$G(y) \triangleq h\mathrm{E}[y-d]^+ + p\mathrm{E}[d-y]^+,$$

where $\mathrm{E}[y-d]^+$ and $\mathrm{E}[d-y]^+$ are the complementary first order loss function and the first order loss function, respectively.

Definition 22. *A replenishment cycle is the time between the placement of two consecutive orders.*

Observe that at the beginning of each replenishment cycle the system "renews" itself, because the inventory position immediately after an order is always equal to S. Let $c(s, S)$ denote the expected long run average cost per period of an (s, S) policy. By using the reward-renewal theorem, Zheng and Federgruen express $c(s, S)$ as the expected cost per cycle divided by the expected cycle length.

For $y > s$, define $k(s, y)$ as the expected total cost until the next order is placed when the starting inventory position is equal to y units. Let $M(j)$ *be the expected total time until an order is placed when starting with* $s + j$ *units*. The expected long run average cost per period of an (s, S) policy can be expressed as

$$c(s,S) = \frac{k(s,S)}{M(S-s)}.$$

We shall next find expressions for $k(s, y)$ and $M(j)$.

Lemma 45. *Consider a discrete random demand d,*

$$c(s,S) = \frac{K + \sum_{j=0}^{S-s-1} m(j)G(S-j)}{M(S-s)},$$

where $M(0) = 0$; $m(0) = (1-p_0)^{-1}$; $m(j) = \sum_{k=0}^{j} p_k m(j-k)$, *for* $j = 1, 2, \ldots$; *and* p_i *is the probability that* $d = i$, *where* $i \geq 0$.[50]

[50] Let $\varphi(k)$ the probability mass function of the demand in a given period; and $\varphi^n(k)$ be the n-fold convolution of $\varphi(k)$. Define $\Phi^n(k) \triangleq \sum_{i=0}^{k} \varphi^n(i)$. Then $M(j) = \sum_{i=1}^{\infty} \Phi^i(j-1) = \sum_{k=0}^{j-1} m(k)$, where $m(j) = \sum_{i=1}^{\infty} \varphi^i(j)$. These results were originally presented in [Veinott and Wagner, 1965].

Proof. Both $k(s, y)$ and $M(j)$ satisfy the discrete renewal equations. Therefore, for $y > s$

$$k(s, y) = G(y) + K \sum_{j=y-s}^{\infty} p_j + \sum_{j=0}^{y-s-1} p_j k(s, y - j)$$

and, for $j = 1, 2, \ldots$

$$M(j) = 1 + \sum_{i=0}^{j-1} p_i M(j - i).$$

From $M(j) = \sum_{k=0}^{j-1} m(k)$ It follows that, for $j = 1, 2, \ldots$

$$M(j) = M(j - 1) + m(j - 1)$$

and therefore, for $y > s$

$$k(s, y) = K + \sum_{j=0}^{y-s-1} m(j) G(y - j).$$

□

$c(s, S)$ is unfortunately not convex. Zheng and Federgruen's efficient algorithm is based on the observation that $c(s-1, S)$ *is a convex combination* of $c(s, S)$ and $G(s)$ (Lemma 46).

Lemma 46.

$$c(s - 1, S) = \alpha_n c(s, S) + (1 - \alpha_n) G(s) \tag{32}$$

where $\alpha_n = M(n)/M(n+1)$.

Proof. For the proof of this and of the following lemmas please refer to [Zheng and Federgruen, 1991]. □

The following corollary immediately follows from Lemma 46.

Corollary 1. $\min\{c(s, S), G(s)\} \leq c(s - 1, S) \leq \max\{c(s, S), G(s)\}$.

By leveraging Lemma 46, one first determines necessary (Lemma 47) and sufficient (Lemma 48) conditions for s^0 to be the optimal reorder-level for a fixed order-up-to level S.

Lemma 47. *If* s^0 *is optimal for* S*, then* $c(s^0, S) \leq c(s, S)$ *for all* s.

Lemma 48. *If* $G(s^0 + 1) \leq c(s^0, S) \leq G(s^0)$*, then* s^0 *is optimal for* S.

The following corollary can be used to find an optimal reorder level for a given order-up-to-level.

Corollary 2. *Given* S*, let* $s^0 \triangleq \max\{y < y_1^* | c(y, S) \leq G(y)\}$*, then Lemma 48 holds, and* s^0 *is optimal for* S.

Let

$$y_1^* \triangleq \min\{y | y = \min_x G(x)\},$$

$$y_2^* \triangleq \max\{y | y = \min_x G(x)\},$$

where $-\infty < y_1^* \leq y^* \leq y_2^* < \infty$ — where y^* is a minimum of $G(x)$.

Next, we establish lower (s_l^*) and upper (s_u^*) bounds that apply to an optimal reorder-level s^*.

Let s_l^* denote the smallest optimal reorder level.

Lemma 49. $s_l^* \leq \bar{s} \triangleq y_1^* - 1$.

Let s_u^* denote the largest optimal reorder level.

Lemma 50. *$s^0 \leq s_u^*$, where s^0 is the optimal reorder-level for some arbitrary order-up-to level S.*

Corollary 3. *There exists an optimal reorder level s^* such that*

$$s^0 \leq s^* \leq \bar{s}.$$

Note that the lower bound can be continuously improved as new optimal reorder level s^0 are found for new values of S.

Next, we establish lower ($\underline{S}$) and upper ($\bar{S}$) bounds that apply to an optimal order-up-to level S^*.

Lemma 51. $\underline{S} \triangleq y_2^*$

Lemma 52. $\bar{S} \triangleq \max\{y > \underline{S} | G(y) < c^*\}$, *where* $c^* = c(s^*, S^*)$.

Corollary 4. *There exists an optimal order-up-to level S^* such that*

$$\underline{S} \leq S^* \leq \bar{S}.$$

Corollary 5. $\bar{S}_c \triangleq \max\{y > \underline{S} | G(y) < c\}$, *where $c = c(s, S)$ is the expected long run average cost per period of an arbitrary (s, S) policy. Therefore, there exists an optimal order-up-to level S^* such that*

$$\underline{S} \leq S^* \leq \bar{S} \leq \bar{S}_c.$$

This corollary can be used to identify increasingly tighter upper bounds for S^* as increasingly better (s, S) policies are found.

For any fixed order up to level S let

$$c^*(S) \triangleq \min_{s<S} c(s, S).$$

Definition 23. *S improves upon S^0, if $c^*(S) < c^*(S^0)$.*

Lemma 53. *For a given order-up-to level S^0 ($\geq y_1^*$), let s_0 ($\leq y_1^*$) be an optimal reorder-level. Then $c^*(S) < c^*(S^0)$ if and only if $c(s^0, S) < c(s^0, S^0)$.*

Thus, given a policy (s^0, S^0), we can easily identify an improving S' by simply comparing $c(s^0, S^0)$ and $c(s^0, S')$. If S' improves upon S^0, then we want to find an optimal reorder-level s' for S'.

Lemma 54. *Assume that $s^0 \leq \bar{s}$ is an optimal reorder-level for S^0, and that S' improves upon S^0, then there exists an optimal reorder-level s' for S' with $s' \in \{s^0, \ldots, \bar{s}\}$.*

The lemma therefore restricts the search for s' to $s^0, \ldots, \bar{s}$.

We next present Zheng and Federgruen's algorithm to find an optimal (s^*, S^*) policy. A sample instance is presented in Listing 73.

```
import numpy as np, functools
from scipy.stats import poisson
from inventoryanalytics.utils import memoize as mem

def expectation(f, x, p): # E[f(X)] = sum f(x_i) p_i
    return np.dot(f(x),p)

class ZhengFedergruen(object):
    """
    The stationary stochastic lot sizing problem.
    """
    def __init__(self, mu, K, h, b):
        """
        Constructs an instance of the stochastic lot sizing problem

        Arguments:
            mu {[type]} -- the expected demand
            K {[type]} -- the fixed ordering cost
            h {[type]} -- the proportional holding cost
            b {[type]} -- the penalty cost
        """

        self.K, self.h, self.b, self.mu = K, h, b, mu
        self.p = poisson.pmf(np.arange(10*mu), self.mu) # set a safe upper bound for
            the random variable support (e.g. 10 times the mean demand)

    def G_L(self, y): # the one-period inventory cost
        return self.b*np.maximum(0,-y) + self.h*np.maximum(0,y)

    def G(self, y): # expected one period inventory cost
        return expectation(self.G_L, y - np.arange(0, len(self.p)), self.p)

    @memoize # see appendix on SDP for memoize class implementation
    def m(self, j): # [Zheng and Federgruen, 1991] eq. 2a
        if j == 0:
            return 1./(1. - self.p[0])
        else: # [Zheng and Federgruen, 1991] eq. 2b
            res = sum(self.p[l]*self.m(j-l) for l in range(1,j+1))
            res /= (1. - self.p[0])
            return res

    @memoize
    def M(self, j): # [Zheng and Federgruen, 1991] eq. 2c
        if j == 0:
            return 0.
        else:
            return self.M(j-1) + self.m(j-1)

    def k(self, s,y): # [Zheng and Federgruen, 1991] eq. 5
        res = self.K
        res += sum(self.m(j)*self.G(y-j) for j in range(y-s))
        return res

    def c(self, s,S): # [Zheng and Federgruen, 1991] eq. 3
        return self.k(s,S)/self.M(S-s)


    def findOptimalPolicy(self):

        # [Zheng and Federgruen, 1991] algorithm on Page 659
        ystar = poisson.ppf(self.b/(self.b+self.h),self.mu).astype(int) #base stock
            level, an arbitrary minimum of G
        s = ystar - 1 # upper bound for s*
        S_0 = ystar + 0 # lower bound for S*

        #calculate the optimal s for S_0 by decreasing s until c(s,S_0) <= G(s)
        while self.c(s,S_0) > self.G(s):
            s -= 1
        s_0 = s # optimal value of s for S_0
        c0 = self.c(s_0,S_0) # expected long run average cost per period of (s_0,S_0)
        S0 = S_0 # S0 = S^0 of the paper
        S = S0 + 1
        while self.G(S) <= c0:
            if self.c(s,S) < c0: # S improves upon S0
                S0 = S
                while self.c(s,S0) <= self.G(s+1): # calculate the optimal s for S_0
                    s += 1
                c0 = self.c(s,S0)
            S += 1
        self.s_star = s
        self.S_star = S0
        return s, S0
```

```
# Optimal policy: (6,40)
# c(s*,S*) = 35.02

instance = {'mu': 10, 'K': 64, 'h':
    1., 'b': 9.}
pb = ZhengFedergruen(**instance)
print(pb.findOptimalPolicy())
```

Listing 73 An instance of the stationary stochastic lot sizing problem. The execution path is illustrated in Fig. 101.

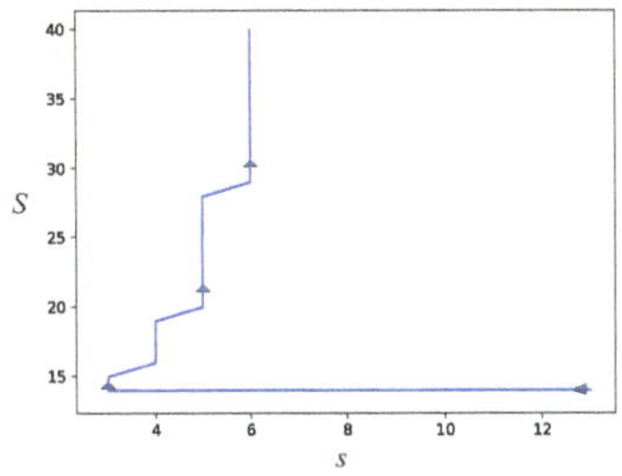

Fig. 101 Execution of the algorithm for the instance in Listing 73.

`findOptimalPolicy`: We enter the algorithm with an initial (arbitrary) order-up-to-level $S^0 = y^*$, where $y^* \triangleq min_y G(y)$. We find an optimal corresponding reorder level s^0 by decreasing s from y^* with unit-sized decrements until $c(s, S^0) <= G(s)$. Optimality of s^0 for S^0 follows from Corollary 2. We then search for the smallest value of S that is larger than S^0 and improves upon S^0. S is increased with unit-sized increments. A single comparison of $c(s^0, S)$ and $c(s^0, S^0)$ is sufficient to verify if a given value of S improves upon S^0 (Lemma 53). If S improves upon S^0, S^0 is updated to S and the new corresponding optimal reorder level s^0 is determined by incrementing the old reorder level with unit-sized increments until $c(s, S^0) > G(s+1)$. The existence of such a reorder level s^0, its optimality for the new value S^0, and the fact that $s_0 < y^*$, are all guaranteed by Lemma 54.

The system can be once more simulated by making a few tweaks to our DES code. In particular, we replace the `OrderUpTo` class with the following `InventoryReview` class.

```
class InventoryReview:
    def __init__(self, des: DES, s: float, S: float, warehouse: Warehouse, lead_time:
            float, B: float):
        self.s, self.S = s, S # the reorder point and the order-up-to-level
        self.w = warehouse # the warehouse
        self.des = des # the Discrete Event Simulation engine
        self.lead_time, self.B = lead_time, B
        self.priority = 1 # denotes a medium priority

    def end(self):
        if self.w.inventory_position() < self.s:
            Q = min(self.S - self.w.inventory_position(), self.B)
            self.w.order(Q, self.des.time)
            self.des.schedule(EventWrapper(ReceiveOrder(self.des, Q, self.w)),
                self.lead_time)
        self.des.schedule(EventWrapper(self), 1) # schedule another review in 1 period
```

Finally, we amend the rest of the code as shown below.

```
np.random.seed(1234)
instance = {"inventory_level": 00, "fixed_ordering_cost": 64, "holding_cost": 1,
     "penalty_cost": 9}
w = Warehouse(**instance)

N = 20 # planning horizon length
des = DES(N)
d = CustomerDemand(des, 10, w)
des.schedule(EventWrapper(d), 0) # schedule a demand immediately

lead_time, capacity, s, S = 0, 10000, 6, 40 # set a very large capacity
o = InventoryReview(des, s, S, w, lead_time, capacity)
des.schedule(EventWrapper(o), 0) # schedule an order immediately
des.schedule(EventWrapper(EndOfPeriod(des, w)), 1) # schedule EndOfPeriod at the
     end of the first period
des.start()

print("Period costs: "+str([w.period_costs[e] for e in w.period_costs]))
print("Average cost per period: "+ '%.2f' % (sum([w.period_costs[e] for e in
     w.period_costs])/len(w.period_costs)))

plot_inventory(w.positions, "inventory position")
plot_inventory(w.levels, "inventory level")
plt.plot([S for k in range(N)], label="S")
plt.plot([s for k in range(N)], label="s")
plt.legend(loc="lower right")
plt.show()
```

The simulated average cost per period (34.8) approaches that computed by Zheng and Federgruen's algorithm (Fig. 102).

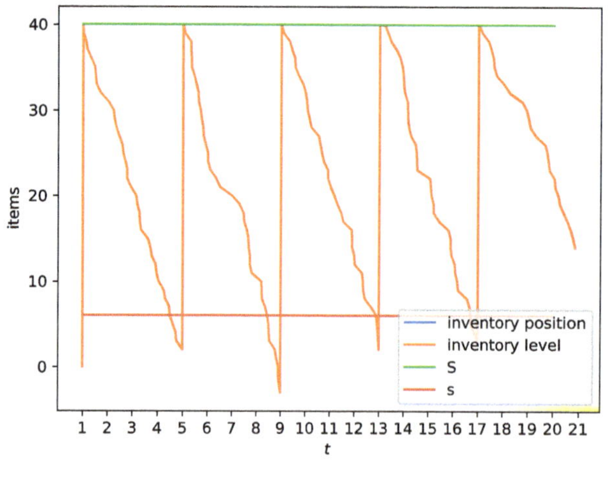

Fig. 102 Simulating the stationary (s, S) policy for the numerical example in Listing 73. If inventory position falls below s at the beginning of any given period, an order is issued to bring the inventory position up to S.

Computing nonstationary (s, S) policy parameters

Definition 24. *A stochastic process $\{d_t\}$ is nonstationary if the distribution of d_t varies with t, i.e. it is time-dependent.*

We present a Stochastic Dynamic Programming algorithm for computing optimal (s_t, S_t) policy parameters for the case in which demand is nonstationary over a finite horizon. An introduction to Stochastic Dynamic Programming is provided in the Appendix.

We first list the relevant imports.

```python
from typing import List
import scipy.stats as sp
import json
```

We implement the state of the inventory system in class `State`.

```python
class State:
    """
    The state of the inventory system.

    Returns:
        [type] -- state of the inventory system
    """

    def __init__(self, t: int, I: float):
        self.t, self.I = t, I # time period t, and inventory level I

    def __eq__(self, other):
        return self.__dict__ == other.__dict__

    def __str__(self):
        return str(self.t) + " " + str(self.I)

    def __hash__(self):
        return hash(str(self))
```

Finally, we implement the Stochastic Dynamic Programming algorithm in class `StochasticLotSizing`

```python
class StochasticLotSizing:
    """
    The nonstationary stochastic lot sizing problem.

    Returns:
        [type] -- A problem instance
    """

    def __init__(self, K: float, v: float, h: float, p: float, d: List[float],
                 max_inv: float, q: float, initial_order: bool):
        """
        Create an instance of StochasticLotSizing.

        Arguments:
            K {float} -- the fixed ordering cost
            v {float} -- the proportional unit ordering cost
            h {float} -- the proportional unit inventory holding cost
            p {float} -- the proportional unit inventory penalty cost
            d {List[float]} -- the demand probability mass function
              taking the form [[d_1,p_1],...,[d_N,p_N]], where d_k is
              the k-th value in the demand support and p_k is its
              probability.
            max_inv {float} -- the maximum inventory level
            q {float} -- quantile truncation for the demand
            initial_order {bool} -- allow order in the first period
        """

        # initialize instance variables
        self.T, self.K, self.v, self.h, self.p, self.d, self.max_inv = len(d)-1, K, v,
            h, p, d, max_inv

        # demand distributions
        max_demand = lambda d: sp.poisson(d).ppf(q).astype(int) # max demand
        pmf = lambda d, k : sp.poisson(d).pmf(k)/q # poisson pmf
        self.pmf = [[[k, pmf(d, k)] for k in range(0, max_demand(d))] for d in self.d]
```

```
        # lambdas
        if initial_order: # action generator
            self.ag = lambda s: [x for x in range(0, max_inv-s.I)]
        else:
            self.ag = lambda s: [x for x in range(0, max_inv-s.I)] if s.t > 0 else [0]
        self.st = lambda s, a, d: State(s.t+1, s.I+a-d) # state transition
        L = lambda i,a,d : self.h*max(i+a-d, 0) + self.p*max(d-i-a, 0) # immediate
            holding/penalty cost
        self.iv = lambda s, a, d: (self.K if a > 0 else 0) + L(s.I, a, d) # immediate
            value function

        self.cache_actions = {} # cache with optimal state/action pairs

    def f(self, level: float) -> float:
        """
        Recursively solve the nonstationary stochastic lot sizing problem
        for an initial inventory level.

        Arguments:
            level {float} -- the initial inventory level

        Returns:
            float -- the cost of an optimal policy
        """
        s = State(0,level)
        return self._f(s)

    def q(self, period: int, level:float) -> float:
        """
        Retrieves the optimal order quantity for a given initial inventory level.
        Function :func:'f' must have been called before using this method.

        Arguments:
            period {int} -- the initial period
            level {float} -- the initial inventory level

        Returns:
            float -- the optimal order quantity
        """
        s = State(period,level)
        return self.cache_actions[str(s)]

    @memoize # see appendix on SDP for memoize class implementation
    def _f(self, s: State) -> float: # Scarf's C_n(x) functional equation
        """
        Dynamic programming forward recursion.

        Arguments:
            s {State} -- the initial state

        Returns:
            float -- the cost of an optimal policy
        """
        #Forward recursion
        v = min(
            [sum([p[1]*(self.iv(s, a, p[0])+ # immediate cost
             (self._f(self.st(s, a, p[0])) if s.t < self.T else 0)) # future cost
                 for p in self.pmf[s.t]]) # demand realisations
             for a in self.ag(s)]) # actions
        opt_a = lambda a: sum([p[1]*(self.iv(s, a, p[0])+
                             (self._f(self.st(s, a, p[0])) if s.t < self.T else 0))
                          for p in self.pmf[s.t]]) == v
        q = [k for k in filter(opt_a, self.ag(s))] # retrieve best action list
        self.cache_actions[str(s)]=q[0] if bool(q) else None # store action
        return v # return expected total cost

    def extract_sS_policy(self) -> List[float]:
        """
        Extract optimal (s,S) policy parameters

        Returns:
            List[float] -- the optimal s,S policy parameters [...,[s_k,S_k],...]
        """
        for i in range(-self.max_inv, self.max_inv):
            self.f(i)
        policy_parameters = []
        for t in range(0, len(self.d)):
            level = self.max_inv - 1
            min_level = -self.max_inv
            s = State(t, level)
            while self.cache_actions.get(str(s), 0) == 0 and level > min_level:
                level, s = level - 1, State(t, level - 1)
            policy_parameters.append(
                [level, level+self.cache_actions.get(str(s), 0)])
        return policy_parameters
```

Example 34. *We consider an instance of the stochastic lot sizing problem over a planning horizon comprising* $N = 4$ *periods. Demand in each period follows a Poisson distribution with mean* λ_t*. The values of* λ_t *are shown in the following table.*

t	1	2	3	4
λ_t	20	40	60	40

The fixed ordering cost is $K = 100$*, for the sake of simplicity the per unit ordering cost is* $v = 0$*, holding cost is* $h = 1$*, and penalty cost is* $p = 10$.

The instance in Example 34 can be solved as follows.

```
instance = {"K": 100, "v": 0, "h": 1, "p": 10, "d": [20,40,60,40], "max_inv": 200,
     "q": 0.9999, "initial_order": True}
lot_sizing = StochasticLotSizing(**instance)
t, I = 0, 0 # initial period, initial inventory level
print("Optimal policy cost: " + str(lot_sizing.f(i)))
print("Optimal order quantity: " + str(lot_sizing.q(t, i)))
print(lot_sizing.extract_sS_policy())
```

The optimal policy cost is 332.1 The optimal (s_t, S_t) policy parameters for each period t are shown in the following table.

t	1	2	3	4
S_t	67	49	109	49
s_t	15	28	55	28

In Fig. 103 we plot Scarf's $G_n(y)$, and $C_n(x)$ functions for the first period of the instance in Example 34. The optimal order quantity Q for each initial inventory level x is also plotted.

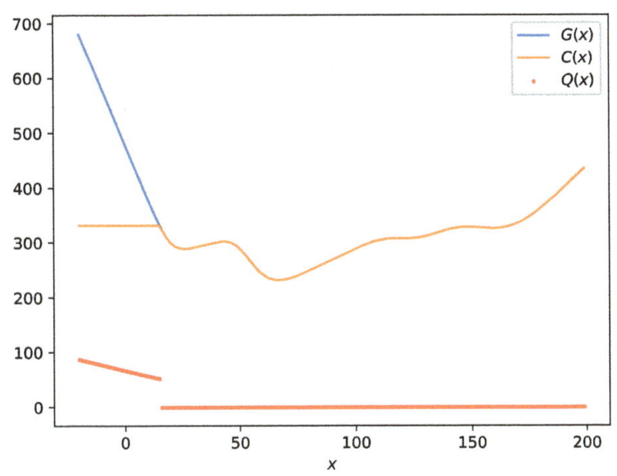

Fig. 103 Scarf's $G_n(y)$ and $C_n(x)$ functions for the first period of the instance in Example 34, i.e. $n = 4$. The optimal order quantity Q for each initial inventory level x is also plotted.

Stochastic Dynamic Programming features a pseudo-polynomial complexity, and can therefore only solve small instances. A more efficient mathematical programming heuristic has been discussed in [Xiang et al., 2018].

The modified (s, S) *policy*

In this section, we consider the same problem addressed in [Scarf, 1960]: inventory of a single item subject to random demand must be controlled over a planning horizon of n periods under fixed ordering (K), holding (h), and backorder penalty (p) cost. However, the manager now also faces order quantity capacity constraints: in each period, the amount ordered cannot exceed a fixed quantity B.

The first to investigate this problem was Wijngaard.[51] Wijngaard conjectured that an optimal strategy may feature a so-called *modified* (s, S) *structure*: if the inventory level is greater or equal to s, do not order; otherwise, order up to S, or as close to S as possible, given the ordering capacity.[52]

[51] Jacob Wijngaard. An inventory problem with constrained order capacity. Technical report, TU Eindhoven, the Netherlands, 1972.

[52] Note that a modified (s, S) policy generally performs very well: it is either optimal, or very close to optimal, and thus it represents a valid heuristic for this problem.

Unfortunately, both [Wijngaard, 1972] and [Shaoxiang and Lambrecht, 1996] provided counterexamples that ruled out the optimality of a modified (s, S) policy. These counterexamples are fairly easy to produce by slightly amending the Stochastic Dynamic Programming code for computing nonstationary (s, S) policy parameters. In particular, the only change that is required concerns the lambda expressions for the action generator

```
if initial_order: # action generator
   self.ag = lambda s: [x for x in range(0, min(max_inv-s.I, B))]
else:
   self.ag = lambda s: [x for x in range(0, min(max_inv-s.I, B)] if s.t > 0 else [0]
```

where `B` is the ordering capacity.[53]

[53] Note that modified (s, S) policy parameters can be extracted by using the function `extract_sS_policy` previously presented.

We hereby illustrate a numerical example originally presented in [Shaoxiang and Lambrecht, 1996, p. 1015] and also investigated in [Shaoxiang, 2004] under an infinite horizon.

Example 35. *Consider a planning horizon of* $n = 20$ *periods and a stationary demand* d *distributed in each period according to the following probability mass function:* $\Pr\{d = 6\} = 0.95$ *and* $\Pr\{d = 7\} = 0.05$. *Other problem parameters are* $K = 22$, $B = 9$, $h = 1$ *and* $p = 10$ *and* $v = 1$*; however, note that for a stationary problem with a sufficiently long horizon,* v *can be safely ignored. The authors also consider a discount factor* $\alpha = 0.9$, *which can be easily embedded in the code we presented.*

In Table 12 we report the tabulated optimal policy as illustrated in [Shaoxiang, 2004, p. 417]. It is easy to see that this policy does not follow a modified (s, S) rule as defined above.

Starting inventory level	-3	-2	-1	0	1	2	3	4	5	6	7
Optimal order quantity	9	8	7	9	8	7	9	8	7	0	0

Table 12 Optimal policy as illustrated in [Shaoxiang, 2004, p. 417].

However, [Shaoxiang and Lambrecht, 1996, Shaoxiang, 2004] provided a partial characterisation of the optimal policy, and proved that the optimal policy for this problem features a so-called $X - Y$ band structure: when initial inventory level is above Y, it is optimal to not order; when initial inventory level is below X, where $Y - X \leq B$, it is optimal to order at full capacity.

Nervousness of control

We consider once more the stochastic lot sizing problem addressed in [Scarf, 1960]. The (s, S) policy is optimal in the sense that it minimises the expected total cost. However, this policy suffers from high *nervousness* of the control action.[54] Tunc et al. distinguish two types of nervousness of the control action: setup-oriented and quantity-oriented. Setup-oriented nervousness is the degree by which timing of next replenishment is unpredictable; quantity-oriented nervousness is the degree by which quantity of next replenishment is unpredictable.[55] Tunc et al. argue that nervousness of control can be categorised in terms of three well-established inventory control strategies: static uncertainty, dynamic uncertainty, and static-dynamic uncertainty, which were originally discussed in [Bookbinder and Tan, 1988], and which reflect extreme cases with regard to the setup- and the quantity-oriented nervousness.

[54] Huseyin Tunc, Onur A. Kilic, S. Armagan Tarim, and Burak Eksioglu. A simple approach for assessing the cost of system nervousness. *International Journal of Production Economics*, 141(2): 619–625, 2013.

[55] While the (s, S) policy is cost optimal, the (R, Q) policy is the most expensive strategy. The cost of static-dynamic uncertainty strategies fall in between these two limiting cases. A comprehensive cost comparison among all these strategies is presented in [Dural-Selcuk et al., 2020, Ma et al., 2020].

THE DYNAMIC UNCERTAINTY STRATEGY is the (s, S) policy discussed in [Scarf, 1960]. In this strategy, at the beginning of each period, the manager reviews the inventory, decides if an order has to be issued, and the size of such order. In this strategy both timing of next replenishment and its quantity are unpredictable, and the strategy therefore features high setup- and quantity-oriented nervousness.

(s, S) policy

THE STATIC-DYNAMIC UNCERTAINTY STRATEGY can be implemented in the form of two policies: the (R, S) policy and the (s, Q) policy.

In the (R, S) policy, at the beginning of the planning horizon, we fix once and for all the length R of all future replenishment cycles; conversely, the size of an order is only computed at the beginning of each replenishment cycle, as the quantity necessary to raise the inventory position up to S. This strategy presents low setup-oriented, and high quantity-oriented nervousness. It is particularly suitable when order coordination across multiple SKUs is a concern, e.g. when multiple SKUs are ordered from the same supplier.

(R, S) policy

In the (s, Q) policy, at the beginning of the planning horizon, we fix once and for all the order quantities for all future periods; conversely, whether an order is issued or not at any given period, is determined by checking if the inventory position has fallen below the reorder point s. The timing of future replenishment periods is therefore uncertain. This strategy presents high setup-oriented nervousness, and low quantity-oriented nervousness. It is particularly suitable when predictability of order processing workload is a concern, e.g. staff rostering for processing orders.

(s, Q) policy

THE STATIC UNCERTAINTY STRATEGY is known as the (R, Q) policy. In this strategy, at the beginning of the planning horizon, we decide once and for all the length R of all future replenishment cycles, and the associated order quantities. This strategy presents low setup-oriented nervousness, and low quantity-oriented nervousness.

(R, Q) policy

The (R,S) policy

We consider the same problem addressed in [Scarf, 1960]: inventory of a single item subject to random demand must be controlled over a planning horizon of n periods under fixed ordering (K), holding (h), and backorder penalty (p) cost. We formulate the problem in Python as follows.

```
from typing import List

class StochasticLotSizing:
    """
    A stochastic lot sizing problem.
    """
    def __init__(self, K: float, h: float, p: float, d: List[float], I0: float):
        """
        Create an instance of the stochastic lot sizing problem.

        Arguments:
            K {float} -- the fixed ordering cost
            h {float} -- the per unit holding cost
            p {float} -- the per unit penalty cost
            d {List[float]} -- the poisson demand rate in each period
            I0 {float} -- the initial inventory level
        """

        self.K, self.h, self.p, self.d, self.I0 = K, h, p, d, I0
```

However, we hereby adopt a static-dynamic uncertainty strategy [Bookbinder and Tan, 1988] in the form of an (R,S) policy. Our aim is to compute near optimal (R,S) policy parameters.

Definition 25. *A replenishment cycle is the time interval $i,\ldots,j$ between two consecutive orders occurring at time i and $j+1$.*

At the beginning of each replenishment cycle of length R, an (R,S) policy raises the inventory position up to S (Fig. 104). Since we operate under an order-up-to policy, the expected total cost $c(i,j)$ of a replenishment cycle spanning over periods $i,\ldots,j$ is the expected total cost of a multi-period Newsvendor problem over the same time interval. The solution to this problem has been already discussed at p. 123. The relevant code is the following.

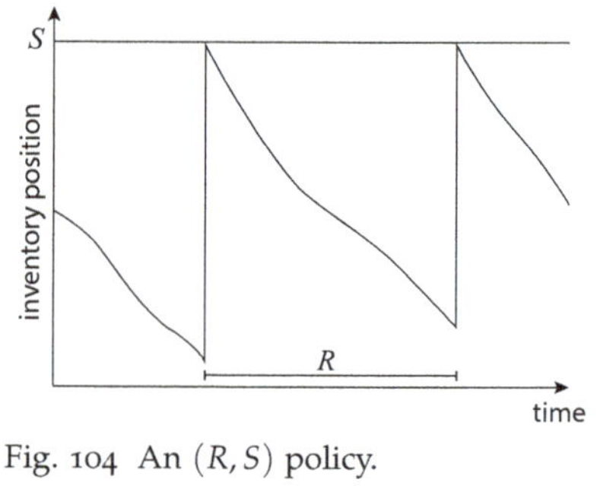

Fig. 104 An (R,S) policy.

```
import scipy.integrate as integrate
from itertools import accumulate
from scipy.stats import poisson
from scipy.optimize import minimize

class MultiPeriodNewsvendor:
    def __init__(self, mean, h, p):
        self.mean, self.o, self.u = mean, h, p

    def cfolf(self, Q, d): # complementary first order loss function
        return integrate.quad(lambda x: poisson.cdf(x, d), 0, Q)[0]

    def folf(self,Q): # first order loss function
        return self.cfolf(Q)-self.u*(Q - self.mean)

    def _C(self, Q): # C(Q)
        return sum([(self.o+self.u)*self.cfolf(Q, d)-self.u*(Q - d) for d in
             accumulate(self.mean)])

    def optC(self): # min C(Q)
        return minimize(self._C, 0, method='Nelder-Mead')
```

Given a planning horizon of n periods, we can precompute the expected total cost $c(i,j)$ of every replenishment cycle, for $i,j \in \{1,\ldots,n\}$. $i<j$.

Once cycle costs have been precomputed, the optimal replenishment plan can be easily determined by adopting the same strategy used to solve the Dynamic Lot Sizing Problem in a deterministic setting (p. 61). In particular, we represent our problem as a Directed Acyclic Graph (DAG) in which arcs represent all possible replenishment cycles that can take place within our n-period planning horizon. Approximating the expected total cost of an optimal plan is equivalent to finding the shortest path in this DAG. This can be done efficiently by leveraging Dijkstra's algorithm [Dijkstra, 1959]. The relevant Python code is shown below and in Listing 74.

```
instance = {"K": 100, "h": 1, "p": 10,
     "d":[20,40,60,40]}
ww = RS_DP(**instance)
print("Cost of an optimal plan: ",
     ww.optimal_cost())
print("Optimal order-up-to-levels: ",
     ww.order_up_to_levels())
```

Listing 74 A stochastic lot sizing problem instance solved under the (R, S) policy. The approximated expected total cost is 388.7. The optimal replenishment plan prescribes orders in periods 1, 3, and 4; the associated order-up-to-levels are 67, 70, 48. Contrast this solution with the optimal (s, S) policy and its cost (332.1) discussed for this example at p. 139.

```
import networkx as nx

class RS_DP(StochasticLotSizing):
    """
    Implements the traditional shortest path algorithm with stochastic cycle costs.

    James H. Bookbinder and Jin-Yan Tan. Strategies for the probabilistic lot-sizing
    problem with service-level constraints. Management Science, 34(9):1096-1108, 1988.
    """

    def __init__(self, K: float, h: float, p: float, d: List[float]):
        """
        Create an instance of the stochastic lot sizing problem.
        Initial inventory level assumed to be 0.

        Arguments:
            K {float} -- the fixed ordering cost
            h {float} -- the per unit holding cost
            p {float} -- the per unit penalty cost
            d {List[float]} -- the demand in each period
        """
        super().__init__(K, h, p, d, 0)

        self.graph = nx.DiGraph()
        for i in range(0, len(self.d)):
            for j in range(i+1, len(self.d)):
                self.graph.add_edge(i, j, weight=self.cycle_cost(i, j-1))

    def cycle_cost(self, i: int, j: int) -> float:
        '''
        Compute the expected total cost of a cycle covering periods i,...,j
        when initial inventory is zero
        '''
        if i>j: raise Exception('i>j')

        return self.K + MultiPeriodNewsvendor(self.d[i:j+1], self.h,self.p).optC().fun

    def optimal_cost(self) -> float:
        '''
        Approximates the cost of an optimal solution
        '''
        T, cost, g = len(self.d), 0, self.graph
        path = nx.dijkstra_path(g, 0, T-1)
        path.append(len(self.d))
        for t in range(1,len(path)):
            cost += self.cycle_cost(path[t-1],path[t]-1)
            print("c("+str(path[t-1])+","+str(path[t]-1)+") =
                "+str(self.cycle_cost(path[t-1],path[t]-1)))
        return cost

    def order_up_to_levels(self) -> List[float]:
        '''
        Compute optimal order-up-to-levels
        '''
        T, g = len(self.d), self.graph
        path = nx.dijkstra_path(g, 0, T-1)
        path.append(len(self.d))
        qty = [0 for k in range(0,T)]
        for t in range(1,len(path)):
            qty[path[t-1]] = MultiPeriodNewsvendor(self.d[path[t-1]:path[t]],
                self.h,self.p).optC().x[0]
        return qty
```

For a stationary demand, the (R, S) policy can be simulated by using a DES code similar to that presented for the (s, S) policy. However, in the revised code, inventory reviews will have to be scheduled in advance, at the timings prescribed by the (R, S) policy.

Advanced mathematical programming heuristics are discussed in [Tarim and Kingsman, 2006, Rossi et al., 2015, Tunc et al., 2018].

The (s, Q) policy

We consider the same problem addressed in [Scarf, 1960]: inventory of a single item subject to random demand must be controlled over a planning horizon of n periods under fixed ordering (K), holding (h), and backorder penalty (p) cost.

We hereby focus on a static-dynamic uncertainty strategy [Bookbinder and Tan, 1988] in the form of an (s, Q) policy. Our aim is to compute near optimal (s, Q) policy parameters.

The DES code required to simulate an (s, Q) policy is essentially identical to that used to simulate an (s, S) policy. The only change required concerns the `InventoryReview` class, which must be amended as follows to capture the fact that when inventory position falls below s, we order a fixed quantity Q.[56]

[56] Recall that in the (s, S) policy the order quantity was dynamically determined to bring the inventory position up to S.

```
class InventoryReview:
   def __init__(self, des: DES, s: float, Q: float, warehouse: Warehouse, lead_time:
        float):
      self.s, self.Q = s, Q # the reorder point and the order quantity
      self.w = warehouse # the warehouse
      self.des = des # the Discrete Event Simulation engine
      self.lead_time = lead_time
      self.priority = 1 # denotes a medium priority

   def end(self):
      if self.w.inventory_position() < self.s:
         self.w.order(self.Q, self.des.time)
         self.des.schedule(EventWrapper(ReceiveOrder(self.des, self.Q, self.w)),
              self.lead_time)
      self.des.schedule(EventWrapper(self), 1) # schedule another review in 1 period
```

Example 36. *Consider the same instance[57] presented in Listing 73. This instance was solved using Zheng and Federgruen's algorithm, which produced an (s, S) policy with parameters $s = 6$ and $S = 40$ and an expected total cost 35.02. We arbitrarily set (s, Q) policy parameters to $s = 6$, and $Q = 40$;[58] and simulate this policy (Fig. 108). The simulated cost (35.6) is slightly higher than of an optimal (s, S) policy.*

[57] Instance parameters: $\mu = 10$, $K = 64$, $h = 1$, and $p = 9$.

[58] Q is thus set to the value of the order-up-to-level under an optimal (s, S) policy.

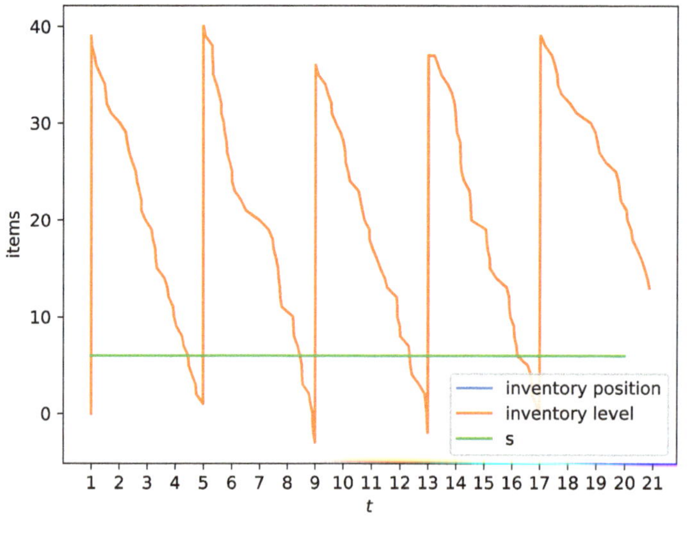

Fig. 105 Simulating the stationary (s, Q) policy for the numerical example in Listing 73. If inventory position falls below s at the beginning of any given period, an order of size Q is issued.

Instead of setting arbitrary (s, Q) policy parameters, we may want to find optimal ones. Unfortunately, computing optimal (s, Q) policy parameters is a challenging task even under stationary demand.[59] We thus approximate this problem numerically using a simulation-optimisation strategy [Jalali and Nieuwenhuyse, 2015].[60]

[59] Advanced mathematical programming heuristics for computing nonstationary (s, Q) policy parameters are discussed in [Ma et al., 2020].

[60] To simulate the cost of a given pair of (s, Q) policy parameters, we leverage our DES code; to optimise over the space of all possible (s, Q) policy parameter pairs we rely on Nelder-Mead's algorithm [Nelder and Mead, 1965].

```
import matplotlib.pyplot as plt, numpy as np, pandas as pd
from scipy.optimize import minimize

class sQ:
   def __init__(self, instance, demand, lead_time, N):
      self.instance, self.demand, self.lead_time, self.N = instance, demand,
           lead_time, N

   def _run_DES(self, parameters):
      s, Q = tuple(parameters)
      np.random.seed(1234)
      w = Warehouse(**self.instance)
      des = DES(self.N)
      d = CustomerDemand(des, self.demand, w)
      des.schedule(EventWrapper(d), 0) # schedule a demand immediately
      o = InventoryReview(des, s, Q, w, self.lead_time)
      des.schedule(EventWrapper(o), 0) # schedule an order immediately
      des.schedule(EventWrapper(EndOfPeriod(des, w)), 1) # schedule EndOfPeriod at
           the end of the first period
      des.start()
      return w

   def simulate(self, parameters):
      w = self._run_DES(parameters)
      return (sum([w.period_costs[e] for e in w.period_costs])/len(w.period_costs))

instance = {"inventory_level": 0, "fixed_ordering_cost": 64, "holding_cost": 1,
     "penalty_cost": 9}
demand, lead_time = 10, 0
N = 1000 # planning horizon length
sq = sQ(instance, demand, lead_time, N)
s, Q = 6, 40 # use optimal (s,S) policy parameters as initial solution for the
     optimisation
m = "Nelder-Mead"
res = minimize(sq.simulate, [s,Q], method=m, options={"maxiter": 50})
print([m, list(res.x), res.fun])
```

Nelder-Mead's algorithm converges to a solution in which $s = 6$, and $Q = 39$;[61] the expected total cost of this (s, Q) policy is 35.06 — only slightly higher than that of an optimal policy. In Fig. 106 we illustrate the execution path of the algorithm.

[61] We have rounded the original solution to the closest integer values since demand follows a Poisson distribution.

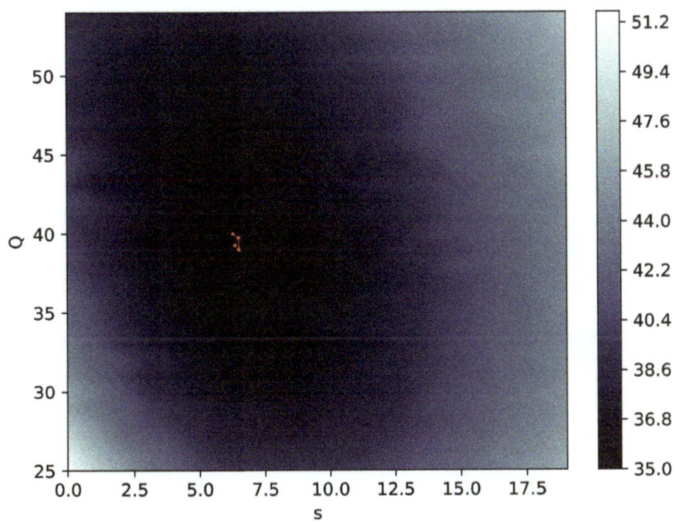

Fig. 106 Nelder-Mead execution path within the landscape of the expected total cost of the (s, Q) policy for the example in Listing 73.

The (R, Q) policy

We consider the same problem addressed in [Scarf, 1960]: inventory of a single item subject to random demand must be controlled over a planning horizon of n periods under fixed ordering (K), holding (h), and backorder penalty (p) cost.

We hereby focus on a static uncertainty strategy [Bookbinder and Tan, 1988] in the form of an (R, Q) policy. Since an (R, Q) policy fixes both the timing and the quantity of replenishments at the beginning of the planning horizon, it can be simulated using the DES code presented at p. 120, when policy parameters are known.

To compute optimal (R, Q) policy parameters, we will formulate the problem using stochastic programming. In particular, we express the model as a mixed-integer nonlinear programming model; and we linearise it using techniques presented in [Rossi et al., 2014b]. The resulting mixed-integer linear programming model can be solved by using off-the-shelf mathematical programming solvers.

Consider a finite planning horizon comprising T periods. Demand d_t is stochastic, and its distribution may vary from one period t to another. We hereby restrict our discussion to a normally distributed demand with mean $\mathrm{E}[d_t]$ and standard deviation $\sigma[d_t]$ in each period t. However the approach here presented can be easily extended to generic distributions. Inventory can only be reviewed — and orders issued — at the beginning of each period. Orders are received immediately after being placed, there is a fixed cost K for placing an order, a per unit cost h for carrying one unit of inventory from one period to the next, and a per unit backorder/penalty cost p for every unit that is backordered at the end of a period. The initial inventory is assumed to be equal to I_0. This problem can be modelled as follows.

$$\min \quad \sum_{t \in T} \delta_t K + h\mathrm{E}[I_t^+] + p\mathrm{E}[I_t^-] \tag{33}$$

Subject to,

$$\delta_t = 0 \rightarrow Q_t = 0 \qquad t = 1, \ldots, T \tag{34}$$

$$I_0 + \sum_{k=0}^{t} (Q_k - \mathrm{E}[d_k]) = \mathrm{E}[I_t] \qquad t = 1, \ldots, T \tag{35}$$

$$\mathrm{E}[I_t] = \mathrm{E}[I_t^+] - \mathrm{E}[I_t^-] \qquad t = 1, \ldots, T \tag{36}$$

$$\mathrm{E}[I_t^+] = \mathrm{E}\left[\sum_{i=1}^{t} (Q_i - d_i)\right]^+ \qquad t = 1, \ldots, T \tag{37}$$

$$\mathrm{E}[I_t^-] = \mathrm{E}\left[\sum_{i=1}^{t} (d_i - Q_i)\right]^+ \qquad t = 1, \ldots, T \tag{38}$$

$$Q_t, \mathrm{E}[I_t^+], \mathrm{E}[I_t^-] \geq 0 \qquad t = 1, \ldots, T \tag{39}$$

where the right hand sides of Eq. 37 and Eq. 38 are the complementary first order loss function (Definition 20) and the first order loss function (Definition 19), respectively.

Following the approach illustrated in [Rossi et al., 2014b], we linearise Eq. 37 and Eq. 38 by means of $W+1$ segments as follows.

$$\mathrm{E}[I_t^+] \geq \mathrm{E}[I_t] \sum_{k=1}^{i} p_k - \sum_{k=1}^{i} p_k \mathrm{E}[\omega_k|\Omega_k]\sigma[d_1+\ldots+d_t] + e_W\sigma[d_1+\ldots+d_t] \quad \text{for } i=1,\ldots,W$$

$$\mathrm{E}[I_t^+] \geq e_W\sigma[d_1+\ldots+d_t]$$

$$\mathrm{E}[I_t^-] \geq -\mathrm{E}[I_t] + \mathrm{E}[I_t] \sum_{k=1}^{i} p_k - \sum_{k=1}^{i} p_k \mathrm{E}[\omega_k|\Omega_k]\sigma[d_1+\ldots+d_t] + e_W\sigma[d_1+\ldots+d_t] \quad \text{for } i=1,\ldots,W$$

$$\mathrm{E}[I_t^+] \geq -\mathrm{E}[I_t] + e_W\sigma[d_1+\ldots+d_t]$$

where p_k, $\mathrm{E}[\omega_k|\Omega_k]$, and e_W are constants that are tabulated in [Rossi et al., 2014b]; a sample linearisation is shown in Fig. 107.

The Python code implementing this mathematical programming model is presented below.

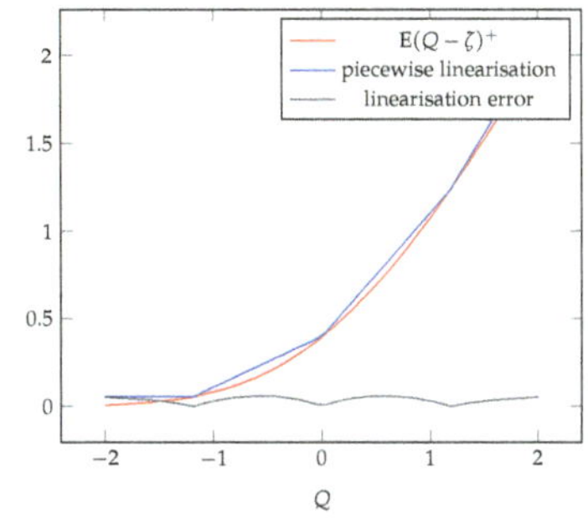

Fig. 107 Edmundson-Madansky [Birge, 2011] (upper) piecewise linearisation (4 segments) of the complementary first order loss function [Rossi et al., 2014b] for a standard normal random variable ζ. Note that the maximum piecewise linearisation error is e_W.

```
import sys
from docplex.mp.model import Model
sys.path.insert(0,'/Applications/CPLEX_Studio1210/cplex/Python/3.7/x86-64_osx')

# http://ibmdecisionoptimization.github.io/docplex-doc/mp/creating_model.html
# http://www-01.ibm.com/support/docview.wss?uid=swg27042869&aid=1

import math, networkx as nx, itertools
from typing import List

class RQ_CPLEX(StochasticLotSizing):
    """
    Solves the RQ problem as an MILP.
    """
    def __init__(self, K: float, h: float, p: float, d: List[float], std_d:
          List[float], I0: float = 0):
        """
        Create an instance of a RQ problem.

        Arguments:
            K {float} -- the fixed ordering cost
            h {float} -- the per unit holding cost
            d {List[float]} -- the demand in each period
            I0 {float} -- the initial inventory level
        """
        super().__init__(K, h, p, d, I0)
        self.std_demand = std_d
        self.W = 5 # 5 partitions, 6 piecewise segments
        # constant linearisation parameters from [Rossi et al., 2014]
        self.prob = [0.1324110437406592, 0.23491250409192982, 0.26535290433482195,
             0.23491250409192987, 0.13241104374065915] # p_k
        self.E = [-1.6180463502161044, -0.6914240068499904, 0, 0.6914240068499903,
             1.6180463502161053] # E[\omega_k|\Omega_k]
        self.e = 0.022270929512393414 # e_W
        self.model()

    def model(self):
        model = Model("RQ")
        T = len(self.d)
        idx = [t for t in range(0,T)]
        self.Q = model.continuous_var_dict(idx, name="Q")
        I = model.continuous_var_dict(idx, name="I") # E[I]
        Ip = model.continuous_var_dict(idx, name="Ip") # E[I^+]
        Im = model.continuous_var_dict(idx, name="Im") # E[I^-]
        delta = model.binary_var_dict(idx, name="delta")

        for t in range(T):
            model.add_constraint(model.if_then(delta[t] == 0, self.Q[t] == 0))
            model.add_constraint(self.I0 + model.sum(self.Q[k] - self.d[k] for k in
                  range(t+1)) == I[t])
            model.add_constraint(I[t] == Ip[t] - Im[t])

            for n in range(self.W): # complementary first order loss function
                model.add_constraint(Ip[t] >= I[t] * sum(self.prob[k] for k in
                      range(n+1)) - sum([self.prob[k]*self.E[k] for k in range(n+1)]) *
                      math.sqrt(sum([self.std_demand[k]**2 for k in range(t+1)])) +
                      self.e * math.sqrt(sum([self.std_demand[k]**2 for k in
                      range(t+1)])))
            model.add_constraint(Ip[t] >= self.e * math.sqrt(sum([self.std_demand[k]**2
                  for k in range(t+1)])))
```

```
            for n in range(self.W): # first order loss function
                model.add_constraint(Im[t] >= -I[t] + I[t] * sum(self.prob[k] for k in
                        range(n+1)) - sum([self.prob[k]*self.E[k] for k in range(n+1)]) *
                        math.sqrt(sum([self.std_demand[k]**2 for k in range(t+1)])) +
                        self.e * math.sqrt(sum([self.std_demand[k]**2 for k in
                        range(t+1)])))
            model.add_constraint(Im[t] >= -I[t] + self.e *
                    math.sqrt(sum([self.std_demand[k]**2 for k in range(t+1)])))

            model.add_constraint(Ip[t] >= 0)
            model.add_constraint(Im[t] >= 0)

        model.minimize(model.sum(delta[t] * self.K + self.h * Ip[t] + self.p * Im[t]
               for t in range(T)))
        model.print_information()
        self.msol = model.solve()
        if self.msol:
            model.print_solution()
        else:
            print("Solve status: " + self.msol.get_solve_status() + "\n")

    def order_quantities(self) -> List[float]:
        return [self.msol.get_var_value(self.Q[t]) for t in range(0,len(self.d))]

    def optimal_cost(self) -> float:
        return self.msol.get_objective_value()

instance = {"K": 300, "h": 1, "p": 20, "d":[100,100,100,100,100,100,100,100],
     "std_d" : [10,10,10,10,10,10,10,10], "I0": 0}
ww = RQ_CPLEX(**instance)
print(ww.order_quantities())
print(ww.optimal_cost())
```

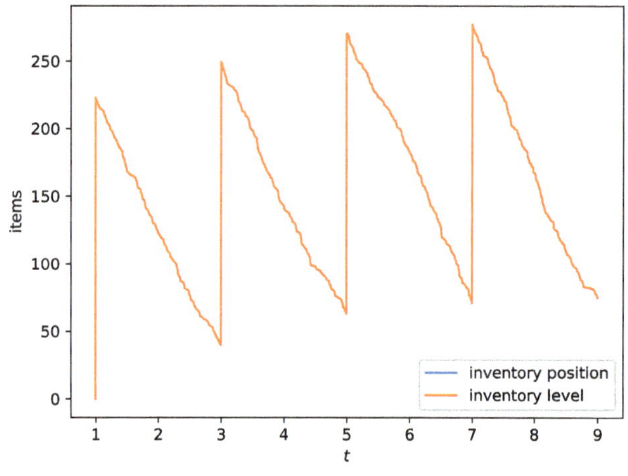

Fig. 108 Simulating the (R, Q) policy for Example 37. The simulated total cost is 2063.04, which is close to that estimated by the mathematical programming model we presented. Observe how the end of period inventory at the end of each replenishment cycle appears to increase as $\sqrt{t}$. If we simulate this system for an infinite number of periods, the expected total cost per period will diverge and go to infinity. An (R, Q) policy therefore **does not control the system** in the long run. To overcome this problem, one can implement a "rolling horizon" control strategy, in which the model is solved for a finite time window, but only the first order quantity Q_1^* in the solution is taken into account and actually implemented (or no order is placed if $Q_1^* = 0$, i.e. if the initial inventory I_0 is sufficient). Then, after a period has passed and demand has been observed, inventory is reviewed, the finite time window is "rolled" by one period, the model is solved again, and the process is repeated. When implemented in the context of a rolling horizon control strategy, the (R, Q) policy becomes a competitive control policy [Dural-Selcuk et al., 2020].

Example 37. *We consider an instance over a planning horizon of $T = 8$ periods. Demand is normally distributed in each period with the following mean and standard deviation.*

t	1	2	3	4	5	6	7	8
μ_t	100	100	100	100	100	100	100	100
σ_t	10	10	10	10	10	10	10	10
Q_t^*	223	0	209	0	207	0	206	0

Other problem parameters are: $K = 300$, $h = 1$, and $p = 20$. We solve the problem using the mathematical programming model presented. The expected total cost of the optimal solution is 1958, the associated order plan (optimal order quantities Q_t^*) is presented in the previous table.

The (R, s, S) policy

We consider the same problem addressed in [Scarf, 1960]: inventory of a single item subject to random demand must be controlled over a planning horizon of n periods under fixed ordering (K), holding (h), and backorder penalty (p) cost. However, we now also include a cost W for reviewing inventory at the beginning of a period.

To control the system, we adopt a hybrid policy that blends the stability of the (R, S) policy and the flexibility of the (s, S) policy. Under an (R, s, S) policy, at the beginning of each replenishment cycle of length R inventory is reviewed at a cost W, if the initial inventory position is below the reorder threshold s, one should issue an order and raise the inventory position up to S (Fig. 109).

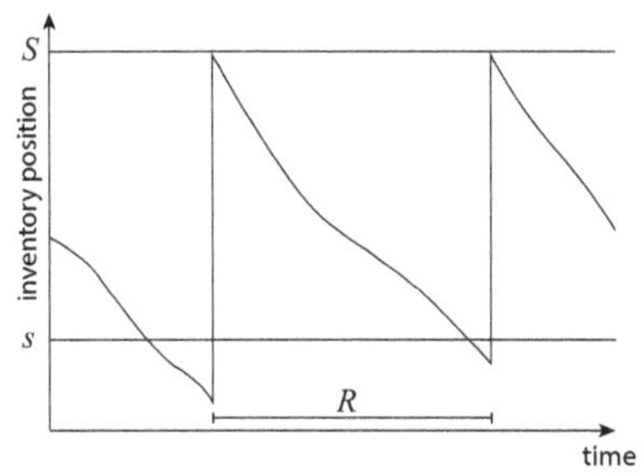

Fig. 109 An (R, s, S) policy.

To compute near optimal (R, s, S) policy parameters, one may rely on a blend of two approaches previously presented [Visentin, 2020]: replenishment cycle lengths R_t and associated order-up-to-positions S_t can be computed by leveraging the Python code in Listing 74, which computes near optimal (R, S) policy parameters; then, for each replenishment cycle, one may determine the associated reorder threshold s by amending the Stochastic Dynamic Programming algorithm for computing optimal (s_t, S_t) policy parameters presented at page 137, so that the order quantity can be non-zero only at the beginning of each replenishment cycle. In both cases, one should set the fixed ordering cost to $K + W$, so that the review cost is taken into account while computing a reorder plan.

Example 38. *We consider an instance of the stochastic lot sizing problem over a planning horizon comprising $N = 4$ periods. Demand in each period follows a Poisson distribution with mean λ_t. The values of λ_t are shown in the following table.*

t	1	2	3	4
λ_t	20	40	60	40

The review cost is $W = 10$, the fixed ordering cost is $K = 100$, for the sake of simplicity the per unit ordering cost is $v = 0$, holding cost is $h = 1$, and penalty cost is $p = 10$.

The optimal (R_t, s_t, S_t) policy cost for the instance presented in Example 38 is 352.3; the optimal (R_t, s_t, S_t) policy parameters for each period t are shown in the following table.

t	1	2	3	4
Review	✓		✓	
S_t	67		109	
s_t	46		86	

A mathematical programming approach for computing (R, s, S) policy parameters is discussed in [Visentin et al., 2021].

Multi-echelon Inventory Systems

https://doi.org/10.11647/OBP.0252.05

Introduction

In this chapter, we briefly survey key aspects related to the control of multi-echelon inventory systems. These are systems in which multiple interconnected installations are present and must be controlled jointly. We first introduce serial systems and associated optimal control strategies; we show how to simulate these systems, and how to compute optimal policy parameters. Finally, we survey other possible multi-echelon inventory systems: assembly systems, distribution systems, and general systems.

Topics

Serial systems

We shall consider a simple serial inventory system comprising two installations: a warehouse W, and a retailer R (Fig. 110). This setup was first investigated in [Clark and Scarf, 1960].

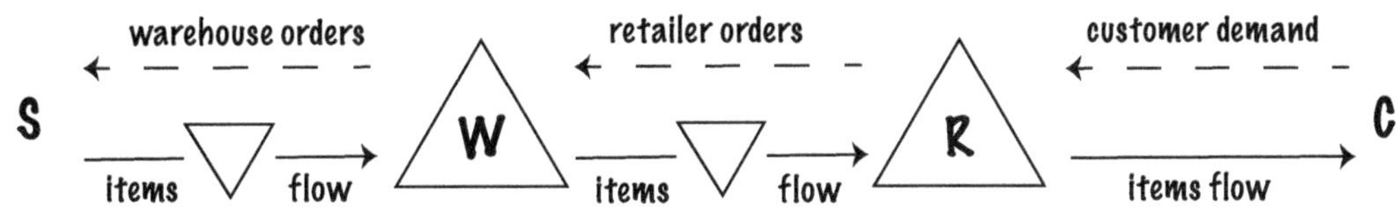

Fig. 110 A serial inventory system comprising two installations: a warehouse and a retailer; physical flows and information flows are represented via solid and dashed lines, respectively.

The system operates in a periodic review setting. We assume installation R faces a customer demand that is stochastic, Poisson distributed with rate λ in each period, and independent across periods. Demand that cannot be met immediately from stock at installation R is backordered. Installation R replenishes from installation W; while installation W replenishes from an infinite outside supply S. Lead times are deterministic and equal to a given (integer) number of periods. There are standard holding costs at installations W and R, and a backorder/penalty cost at installation R. For the time being, we will assume that no ordering/setup costs are present.

At the beginning of each period, the order of events is as follows:

- installation W orders;
- the period delivery from the outside supplier S arrives at W;
- installation R orders from installation W;
- the period delivery from installation W arrives at installation R;
- the stochastic customer demand at installation R is realised;
- evaluation of holding and shortage costs.

We introduce the following notation:

l_i	lead time at installation $i \in \{W, R\}$;
h_i	holding cost per period at installation $i \in \{W, R\}$;
b	backorder/penalty cost per period (only charged at installation R);
I_i	installation stock inventory level at installation $i \in \{W, R\}$ just before period demand;
d_n	n-period stochastic demand, i.e. Poisson($n\lambda$).

Our aim is to minimise the expected total holding and backordering cost per period.

Note that we only charge holding costs for stock on hand at the installation, i.e. we do not charge holding costs on in-transit stock between installation W and installation R, as it can be proved that this cost ($h_W \lambda l_W$) is not affected by the control policy.

Simulating a serial system

We will simulate a simple serial inventory system comprising two installations via DES. We first introduce our usual DES engine class.

```
import matplotlib.pyplot as plt, numpy as np, pandas as pd
from queue import PriorityQueue
from collections import defaultdict
from typing import List

class EventWrapper():
   def __init__(self, event):
      self.event = event

   def __lt__(self, other):
      return self.event.priority < other.event.priority

class DES():
   def __init__(self, end):
      self.events, self.end, self.time = PriorityQueue() , end, 0

   def start(self):
      while True:
         event = self.events.get()
         self.time = event[0]
         if self.time <= self.end:
            event[1].event.end()
         else:
            break

   def schedule(self, event: EventWrapper, time_lag: int):
      self.events.put((self.time + time_lag, event))
```

Next, we model our two installations: the warehouse W,

```
class Warehouse:
   def __init__(self, inventory_level, holding_cost, lead_time):
      self.i, self.h, self.lead_time = inventory_level, holding_cost, lead_time
      self.o = 0 # outstanding_orders
      self.period_costs = defaultdict(int) # a dictionary recording cost in each
           period

   def receive_order(self, Q, time):
      self.review_inventory(time)
      self.i, self.o = self.i + Q, self.o - Q
      self.review_inventory(time)

   def order(self, Q, time):
      self.review_inventory(time)
      self.o += Q
      self.review_inventory(time)

   def on_hand_inventory(self):
      return max(0,self.i)

   def backorders(self):
      return max(0,-self.i)

   def issue(self, demand, time):
      self.review_inventory(time)
      self.i = self.i-demand

   def inventory_position(self):
      return self.o+self.i

   def review_inventory(self, time):
      try:
         self.levels.append([time, self.i])
         self.on_hand.append([time, self.on_hand_inventory()])
         self.positions.append([time, self.inventory_position()])
      except AttributeError:
         self.levels, self.on_hand = [[0, self.i]], [[0, self.on_hand_inventory()]]
         self.positions = [[0, self.inventory_position()]]

   def incur_end_of_period_costs(self, time): # incur holding and penalty costs
      self._incur_holding_cost(time)

   def _incur_holding_cost(self, time): # incur holding cost and store it in a
        dictionary
      self.period_costs[time] += self.on_hand_inventory()*self.h
```

and the retailer R.

```
class Retailer:
    def __init__(self, inventory_level, holding_cost, penalty_cost, lead_time,
            demand_rate):
        self.i, self.h, self.p, self.lead_time, self.demand_rate = inventory_level,
            holding_cost, penalty_cost, lead_time, demand_rate
        self.o = 0 # outstanding_orders
        self.period_costs = defaultdict(int) # a dictionary recording cost in each
            period

    def receive_order(self, Q, time):
        self.review_inventory(time)
        self.i, self.o = self.i + Q, self.o - Q
        self.review_inventory(time)

    def order(self, Q, time):
        self.review_inventory(time)
        self.o += Q
        self.review_inventory(time)

    def on_hand_inventory(self):
        return max(0,self.i)

    def backorders(self):
        return max(0,-self.i)

    def issue(self, demand, time):
        self.review_inventory(time)
        self.i = self.i-demand

    def inventory_position(self):
        return self.o+self.i

    def review_inventory(self, time):
        try:
            self.levels.append([time, self.i])
            self.on_hand.append([time, self.on_hand_inventory()])
            self.positions.append([time, self.inventory_position()])
        except AttributeError:
            self.levels, self.on_hand = [[0, self.i]], [[0, self.on_hand_inventory()]]
            self.positions = [[0, self.inventory_position()]]

    def incur_end_of_period_costs(self, time): # incur holding and penalty costs
        self._incur_holding_cost(time)
        self._incur_penalty_cost(time)

    def _incur_holding_cost(self, time): # incur holding cost and store it in a
            dictionary
        self.period_costs[time] += self.on_hand_inventory()*self.h

    def _incur_penalty_cost(self, time): # incur penalty cost and store it in a
            dictionary
        self.period_costs[time] += self.backorders()*self.p
```

Finally, we model the relevant events, to which we assign priorities in line with the order previously illustrated.

```
class CustomerDemand:
    def __init__(self, des: DES, demand_rate: float, retailer: Retailer):
        self.d = demand_rate # the demand rate per period
        self.r = retailer # the retailer
        self.des = des # the Discrete Event Simulation engine
        self.priority = 5 # denotes a low priority

    def end(self):
        self.r.issue(1, self.des.time)
        self.des.schedule(EventWrapper(self), np.random.exponential(1/self.d)) #
            schedule another demand

class EndOfPeriod:
    def __init__(self, des: DES, warehouse: Warehouse, retailer: Retailer):
        self.w = warehouse # the warehouse
        self.r = retailer # the retailer
        self.des = des # the Discrete Event Simulation engine
        self.priority = 0 # denotes a high priority

    def end(self):
        self.w.incur_end_of_period_costs(self.des.time-1)
        self.r.incur_end_of_period_costs(self.des.time-1)
        self.des.schedule(EventWrapper(EndOfPeriod(self.des, self.w, self.r)), 1)
```

```python
class OrderUpTo_Warehouse:
    def __init__(self, des: DES, S: float, warehouse: Warehouse, retailer: Retailer):
        self.S = S # the order-up-to-position
        self.w = warehouse # the warehouse
        self.r = retailer # the retailer
        self.des = des # the Discrete Event Simulation engine
        self.priority = 1 # denotes a medium priority

    def end(self):
        Q = self.S - self.w.inventory_position()
        self.w.order(Q, self.des.time)
        self.des.schedule(EventWrapper(ReceiveOrder_Warehouse(self.des, Q, self.w,
            self.r)), self.w.lead_time)

class OrderUpTo_Retailer:
    def __init__(self, des: DES, S: float, warehouse: Warehouse, retailer: Retailer):
        self.S = S # the order-up-to-position
        self.w = warehouse # the warehouse
        self.r = retailer # the retailer
        self.des = des # the Discrete Event Simulation engine
        self.priority = 2 # denotes a medium priority

    def end(self):
        Q = self.S - self.r.inventory_position()
        self.r.order(Q, self.des.time)
        Q_available = min(Q, self.w.on_hand_inventory())
        self.w.issue(Q, self.des.time)
        self.des.schedule(EventWrapper(ReceiveOrder_Retailer(self.des, Q_available,
            self.r)), self.r.lead_time)

class ReceiveOrder_Warehouse:
    def __init__(self, des: DES, Q: float, warehouse: Warehouse, retailer: Retailer):
        self.Q = Q # the order quantity
        self.w = warehouse # the warehouse
        self.r = retailer # the retailer
        self.des = des # the Discrete Event Simulation engine
        self.priority = 3 # denotes a medium priority

    def end(self):
        backorders = self.w.backorders()
        self.w.receive_order(self.Q, self.des.time)
        if backorders > 0:
            q = min(self.Q, backorders)
            self.des.schedule(EventWrapper(ReceiveOrder_Retailer(self.des, q, self.r)),
                self.r.lead_time)

class ReceiveOrder_Retailer:
    def __init__(self, des: DES, Q: float, retailer: Retailer):
        self.Q = Q # the order quantity
        self.r = retailer # the retailer
        self.des = des # the Discrete Event Simulation engine
        self.priority = 4 # denotes a medium priority

    def end(self):
        self.r.receive_order(self.Q, self.des.time)
```

The `simulate` method runs the DES (Fig. 111). `S_r` represents the order up to position for the retailer, and `S_w` represents the order up to position for the warehouse, `N` is the number of periods to be simulated.

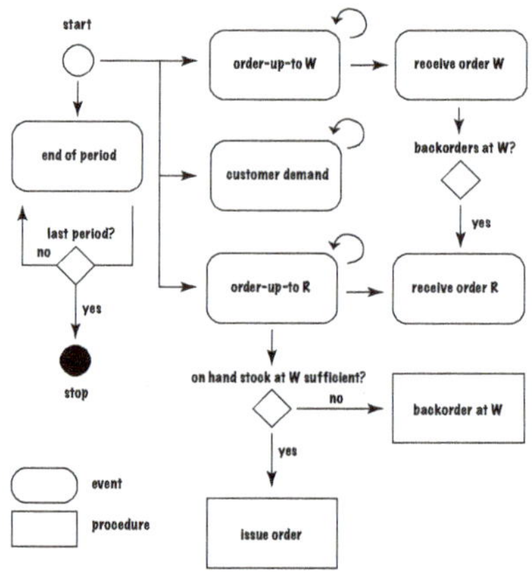

Fig. 111 A serial inventory system comprising two installations: DES flow diagram.

```python
def simulate(retailer, S_r, warehouse, S_w, N, plot):
    np.random.seed(1234)
    r, w = Retailer(**retailer), Warehouse(**warehouse)

    des = DES(N)
    d = CustomerDemand(des, r.demand_rate, r)
    des.schedule(EventWrapper(d), 0) # schedule a demand immediately

    o_r = OrderUpTo_Retailer(des, S_r, w, r)
    o_w = OrderUpTo_Warehouse(des, S_w, w, r)
    for t in range(N):
        des.schedule(EventWrapper(o_r), t) # schedule orders
        des.schedule(EventWrapper(o_w), t) # schedule orders
    des.schedule(EventWrapper(EndOfPeriod(des, w, r)), 1) # schedule EndOfPeriod at
        the end of the first period
    des.start()

    tc = sum([w.period_costs[e] for e in w.period_costs]) + sum([r.period_costs[e]
        for e in r.period_costs])
    return tc/N
```

Example 39. *We consider the following instance: retailer holding cost* $h_r = 1.5$*, retailer backorder/penalty cost* $b = 10$*, retailer order leadtime* $l_r = 5$*, customer demand follows a Poisson distribution with rate* $\lambda = 10$*, warehouse holding cost* $h_w = 1$*, warehouse leadtime* $l_w = 5$*. For the sake of illustration we will set* $S_r = 74$ *and* $S_w = 59$*.*

The previous example can be simulated as follows.

```
N = 10000 # planning horizon length
S_r, S_w = 74, 59
retailer = {"inventory_level": S_r, "holding_cost": 1.5, "penalty_cost": 10,
     "lead_time": 5, "demand_rate": 10}
warehouse = {"inventory_level": S_w, "holding_cost": 1, "lead_time": 5}
print("Avg cost per period: "+ '%.2f' % simulate(retailer, S_r, warehouse, S_w, N))
```

The simulated average cost per period is 26.11. The behaviour of the inventory level for periods $5, \ldots, 14$ is shown in Fig. 112.

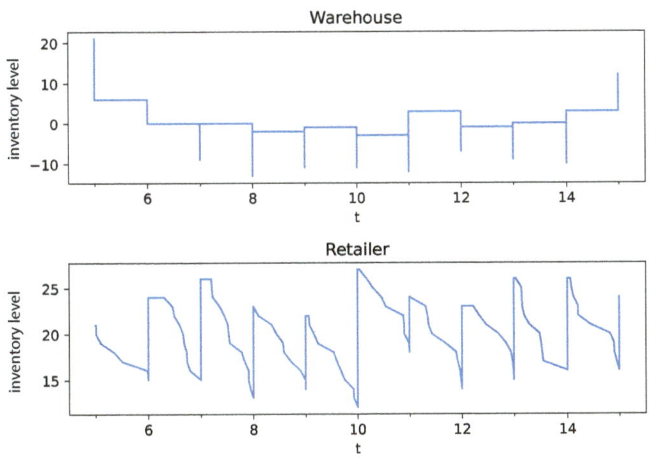

Fig. 112 A serial inventory system comprising two installations: behaviour of the inventory level at installations W and R.

Order-up-to-positions at the retailer and warehouse have been set arbitrarily to $S_r = 74$ and $S_w = 59$, respectively. However, a manger would ideally like to set optimal values for S_r and S_w. A naïve approach to computing optimal values for S_r and S_w is the brute force approach, which explores all possible integer[62] combinations of S_r and S_w (Fig. 113).

[62] Since demand can only take integer values, we can restrict the search to integer combinations.

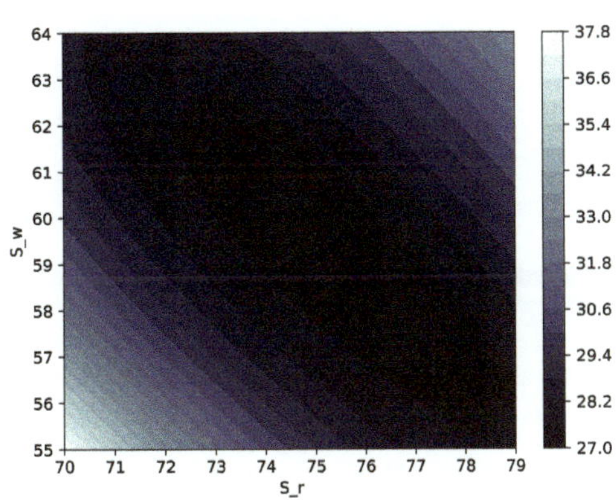

Fig. 113 A serial inventory system comprising two installations: average cost per period for different combination of S_r and S_w; the chosen combination $S_r = 74$ and $S_w = 59$ appears to minimise the expected total cost per period, or at least to be a solution close to the optimal one. Observe that the cost function appears to be convex.

Computing optimal base-stock policy parameters

Instead of relying on the brute force approach, we here illustrate an exact approach[63] for computing optimal base-stock policy parameters for serial inventory systems under stationary demand and an infinite planning horizon. Without loss of generality, we shall focus on the two-installations case; these results are easily extended to an arbitrary number of installations.

[63] Andrew J. Clark and Herbert Scarf. Optimal policies for a multi-echelon inventory problem. *Management Science*, 6(4):475–490, 1960.

Let O_i denote the **outstanding orders**[64] at installation i in any given period; and recall the following definitions.

[64] Outstanding orders are orders that have been issued but not yet received due to the delivery lead time.

The **on hand inventory** at installation i is the positive part of the installation stock inventory level I_i, i.e. $\max\{I_i, 0\} \triangleq [I_i]^+$.

The **installation stock inventory position** Y_i is defined as follows: $Y_i \triangleq O_i + I_i$.

In their exact approach, Clark and Scarf leverage the concept of **echelon stock** (Fig. 114) to compute optimal base-stock policy parameters for serial inventory systems.

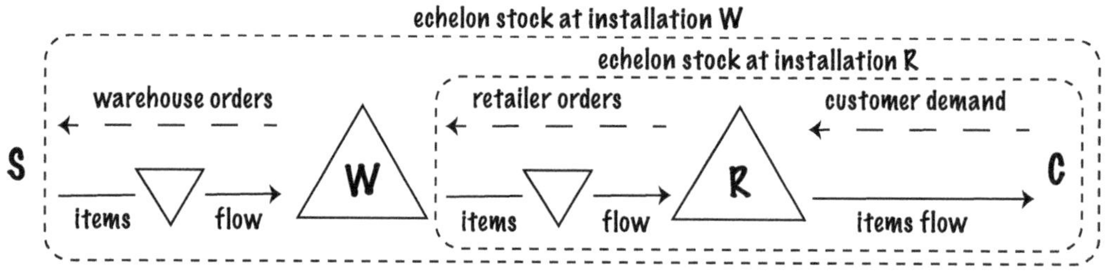

Fig. 114 A serial inventory system comprising two installations: echelon stock at installations W and R.

In essence, *an echelon stock tracks not just the installation stock, but also the downstream stock of an item.*

For an installation i that serves other downstream installations,[65] the **echelon stock inventory position** Y_i^e is defined as follows.

[65] Such as our installation W, the warehouse.

Definition 26. *The echelon stock inventory position is the installation stock inventory position plus the sum of the installation inventory positions at all downstream installations.*

Hence, e.g. $Y_W^e \triangleq Y_W + Y_R$.

Corollary 6. *For installations that do not have further downstream installations they serve,*[66] *the echelon stock inventory position is the same as the installation stock inventory position.*

[66] Such as our installation R, the retailer.

Finally, the **realised echelon stock inventory position** is defined as follows.

Definition 27. *The realised echelon stock inventory position y_i^e at installation i is equal to the echelon stock inventory position Y_i^e minus those outstanding orders, which are backordered at the installation upstream.*[67]

[67] Observe that if the retailer issues an order of size Q at time t, receipt of such a quantity at time $t + l_R$ is not guaranteed, because the warehouse may have backordered this request due to lack of sufficient on hand stock.

Clearly, since the warehouse W replenishes from an infinite supply S, its realised echelon stock inventory position is equal to its echelon stock inventory position. While at the retailer R, these two quantities may differ.

Recall that our aim is to *minimise the expected total holding and backordering cost per period*; and that we only charge holding costs for stock on hand at the installation, i.e. we do not charge holding costs on in-transit stock between installation W and installation R.

We define the **echelon holding cost** $e_i = h_i - h_{i-1}$ at installation i, as the holding cost on the value added when going from installation $i-1$ to installation i, where $i-1$ denotes the installation *upstream* to installation i. Clearly, $e_W = h_W$ since the warehouse has no upstream installation; while $e_R = h_R - h_W$.

First, we should bear in mind that, since the system is stationary, all periods are equal for the purpose of computing the expected total cost per period.

For installation W we consider an order in period t, and costs at the end of period $t + l_W$. For installation R, we consider an order in period $t + l_W$, and costs at the end of period $t + l_W + l_R$.

Lemma 55. *Consider period $t + l_W$, the installation stock inventory on hand at W, after serving the order received from installation R, is*

$$[I_W]^+ = I^e_W - y^e_R$$

and remains stable throughout the rest of the period.

Lemma 56. *The average holding costs at installation W in period $t + l_W$ is*

$$\begin{aligned} h_W E[I^e_W - y^e_R] &= h_W E[y^e_W - d_{l_W+1} - y^e_R] \\ &= h_W(y^e_W - (l_W+1)\lambda) - h_W y^e_R \end{aligned} \tag{40}$$

Lemma 57. *The average holding and backordering costs at installation R in period $t + l_W + l_R$ are*

$$\begin{aligned} &h_R E[y^e_R - d_{l_R+1}]^+ + bE[d_{l_R+1} - y^e_R]^+ \\ &= h_R(y^e_R - (l_R+1)\lambda) + (h_R + b)E[d_{l_R+1} - y^e_R]^+ \end{aligned} \tag{41}$$

Observe that the expected total period costs are a function of y^e_R and y^e_W.

We now reallocate term $-h_W y^e_R$ from Eq. 40 to Eq. 41. By using the fact that $h_R - h_W = e_R$, we obtain

$$C_W(y^e_W) \triangleq h_W(y^e_W - (l_W+1)\lambda) \tag{42}$$

$$C_R(y^e_R) \triangleq e_R y^e_R - h_R(l_R+1)\lambda + (h_R+b)\mathrm{E}[d_{l_R+1} - y^e_R]^+ \tag{43}$$

where Eq. 42 is independent of y^e_R, and Eq. 43 is independent of y^e_W.

Observe that $\mathcal{L}(y) \triangleq \mathrm{E}[d_{l_R+1} - y]^+$ is the first order loss function, which is convex; and since $\mathcal{L}'(y) = F(y) - 1$, where F is the cumulative distribution of d_{l_R+1} [Rossi et al., 2014b, Lemma 1], then the optimal $\hat{y}^e_R$ can be easily obtained from the first order condition

$$\frac{\mathrm{d}C_R(y)}{\mathrm{d}y} = e_R + (h_R + b)(F(y) - 1) = 0$$

that is, by solving

$$F(y) = \frac{h_W + b}{h_R + b}.$$

From the definition of y^e_W and y^e_R, it follows that

$$y^e_R \leq I^e_W = y^e_W - d_{l_W+1}.$$

If $y^e_W - d_{l_W+1} \geq \hat{y}^e_R$, then $\hat{y}^e_R$ is the value that minimises C_R. However, if $y^e_W - d_{l_W+1} < \hat{y}^e_R$, then $\hat{y}^e_R$ cannot be attained, and the best possible value that minimises $C_R(y)$ is $y = y^e_W - d_{l_W+1}$, due to convexity of C_R. This means that the optimal policy is an (echelon stock) base-stock policy with (echelon) order-up-to-position $S^e_R = \hat{y}^e_R$. Furthermore, this optimal policy at R is independent of y^e_W.

Finally, we determine the optimal policy at W. We consider the expected total cost C for the system, when an optimal policy is implemented at R:

$$C(y^e_W) \triangleq C_W(y^e_W) + C_R(\hat{y}^e_R) + \underbrace{\sum_{y^e_W - \hat{y}^e_R}^{\infty} (C_R(y^e_W - u) - C_R(\hat{y}^e_R)) f(u)}_{\text{shortage costs at installation } W},$$

where f is the probability mass function of d_{l_W+1}, that is a Poisson distributed random variable with rate $(l_W + 1)\lambda$. The last term of C can be interpreted as the shortage costs at installation W induced by its inability to deliver on time to installation R.

Function C is convex, and therefore can be easily minimised. Let $\hat{y}^e_W$ be the global minimum of C, since supplier S has infinite capacity, the optimal policy at W is an (echelon stock) base-stock policy with (echelon) order-up-to-position $S^e_W = \hat{y}^e_W$.

Finally, observe that it is easy to switch from an echelon to an installation base-stock policy, by bearing in mind that the installation order-up-to-position $S_W = S^e_W - S^e_R$.

We next present a Python implementation of these results.

```
import math, matplotlib.pyplot as plt
from scipy.stats import poisson

# cost at the retailer
def C_R(y, e_R, h_R, b, L_R, demand_rate):
   M = round(6*math.sqrt((L_R+1)*demand_rate)+(L_R+1)*demand_rate) # safe upper
       bound: 6 sigma
   return y*e_R-h_R*(L_R+1)*demand_rate+(h_R+b)*sum([(d-y)*poisson.pmf(d,
        (L_R+1)*demand_rate) for d in range(y,M)])

# retailer (echelon) order-up-to-position
def compute_y_R(h_W, h_R, b, L_R, demand_rate):
   return poisson.ppf((h_W + b)/(h_R + b), (L_R+1)*demand_rate)

# expected total cost C for the serial system, when an optimal policy is
       implemented at R
def C(y, e_R, h_R, b, L_R, h_W, L_W, demand_rate):
   y_R = int(compute_y_R(h_W, h_R, b, L_R, demand_rate))
   CW = h_W*(y - (L_W+1)*demand_rate)
   CR = C_R(y_R, e_R, h_R, b, L_R, demand_rate)
   M = round(6*math.sqrt((L_W+1)*demand_rate)+(L_W+1)*demand_rate)
   s = sum([(C_R(y-d, e_R, h_R, b, L_R, demand_rate) - CR)*poisson.pmf(d,
        (L_W+1)*demand_rate) for d in range(y-y_R,M)])
   return CW + CR + s

# warehouse (echelon) order-up-to-position
def compute_y_W(e_R, h_R, b, L_R, h_W, L_W, demand_rate, initial_value):
   y, c = initial_value, C(initial_value, e_R, h_R, b, L_R, h_W, L_W, demand_rate)
   c_new = C(y + 1, e_R, h_R, b, L_R, h_W, L_W, demand_rate)
   while c_new < c:
      c = c_new
      y = y + 1
      c_new = C(y + 1, e_R, h_R, b, L_R, h_W, L_W, demand_rate)
   return y
```

We consider once more the instance in Example 39. The optimal solution can be obtained as follows.

```
retailer = {"holding_cost": 1.5, "penalty_cost": 10, "lead_time": 5, "demand_rate":
     10}
warehouse = {"holding_cost": 1, "lead_time": 5}

h_W, h_R = warehouse["holding_cost"], retailer["holding_cost"]
e_W = h_W
e_R = h_R - e_W
b, demand_rate = retailer["penalty_cost"], retailer["demand_rate"]
L_R, L_W = retailer["lead_time"], warehouse["lead_time"]

initial_value = 100
ye_R = compute_y_R(h_W, h_R, b, L_R, demand_rate)
ye_W = compute_y_W(e_R, h_R, b, L_R, h_W, L_W, demand_rate, initial_value)
print("y^e_R="+str(ye_R))
print("y^e_W="+str(ye_W))
print("y_W="+str(ye_W-ye_R))
print("C(y^e_W)="+str(C(ye_W, e_R, h_R, b, L_R, h_W, L_W, demand_rate)))
```

The solution is $\hat{y}^e_R = S^e_R = S_R = 74$, $\hat{y}^e_W = S^e_W = 133$, $\hat{y}_W = S^e_W - S^e_R = S_w = 59$; and has a cost $C(\hat{y}^e_W) = 26.36$. This is indeed the same solution we previously considered. In Fig. 115 we plot $C_R(y)$; in Fig. 116 we plot the expected total cost $C(y)$ for the system.

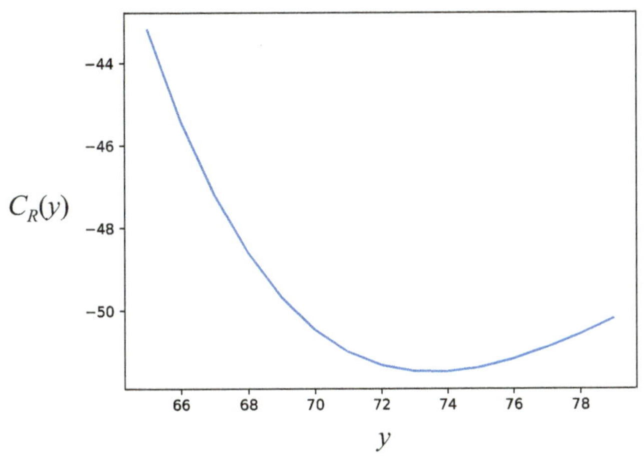

Fig. 115 A serial inventory system comprising two installations: $C_R(y)$.

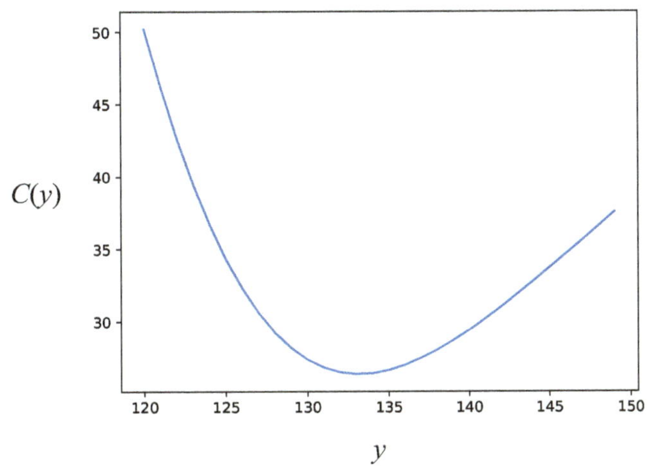

Fig. 116 A serial inventory system comprising two installations: $C(y)$.

Assembly systems

An assembly system is a multi-echelon production line in which end products are manufactured from more basic components. An example of an assembly system is shown in Fig. 117.

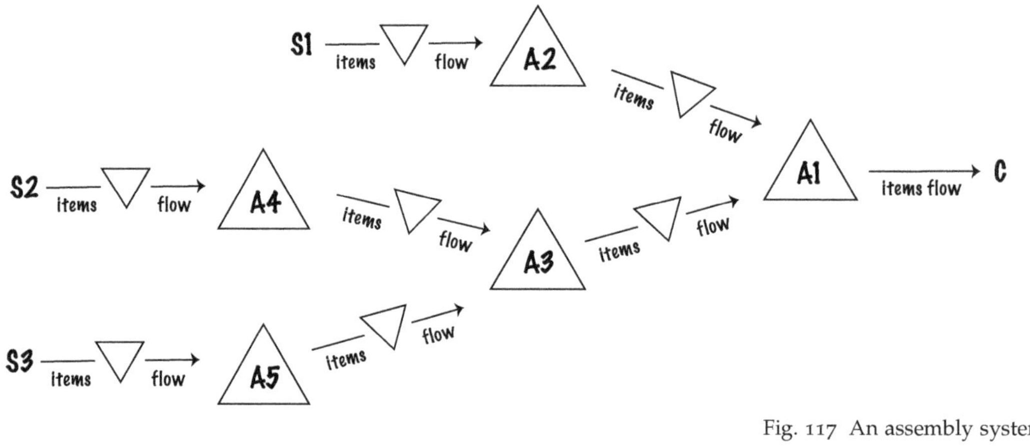

Fig. 117 An assembly system.

Every assembly system can be transformed into an equivalent serial system,[68] therefore dealing with assembly systems is no more difficult than dealing with its equivalent serial system.

[68] Kaj Rosling. Optimal inventory policies for assembly systems under random demands. *Operations Research*, 37(4):565–579, 1989.

Distribution systems

A distribution system is a multi-echelon inventory system in which a given product is distributed from a supply to a number of downstream installations, in order to serve some sources of demand. An example of a distribution system is shown in Fig. 118.

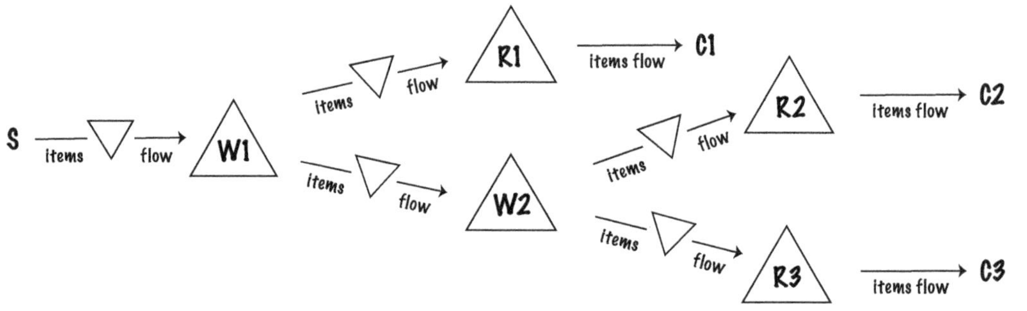

Fig. 118 A distribution system.

Distribution systems are difficult to control. The difficulty stems from the fact that, when inventory is scarce, upstream installations must decide how to allocate this scarce resource optimally to serve demand coming from their downstream installations. Possible stock allocation strategies may include: first-come first-serve, stock balancing, or priority-based. The structure of the optimal control policy is generally not known for these systems; it is therefore customary to fix a control policy (e.g. base-stock policy) and an allocation strategy (e.g. first-come first-serve), and then compute optimal or near-optimal policy parameters under these assumptions.

A well-known distribution system that has been widely studied in the literature is the one-warehouse multiple-retailers system, which comprises a single upstream installation (the warehouse) that serves directly several downstream installations (the retailers). It is common to use Clark and Scarf's approach for tackling this system. However, this approach is no longer exact. The underpinning approximation consists in allowing the warehouse to implement negative allocations at its downstream installations. This means the total stock at the retailers can be optimally "reshuffled" between sites at any period. This technique was illustrated in [Clark and Scarf, 1960]. A well-known heuristic approach for this system is the so-called METRIC, which was originally introduced in [Sherbrooke, 1968]. A computationally expensive exact approach, the "projection algorithm," to tackle the one-warehouse multiple-retailers system was introduced in [Axsäter, 1990].

The one-warehouse multiple-retailers system

General systems

General systems take the form illustrated in Fig. 119.

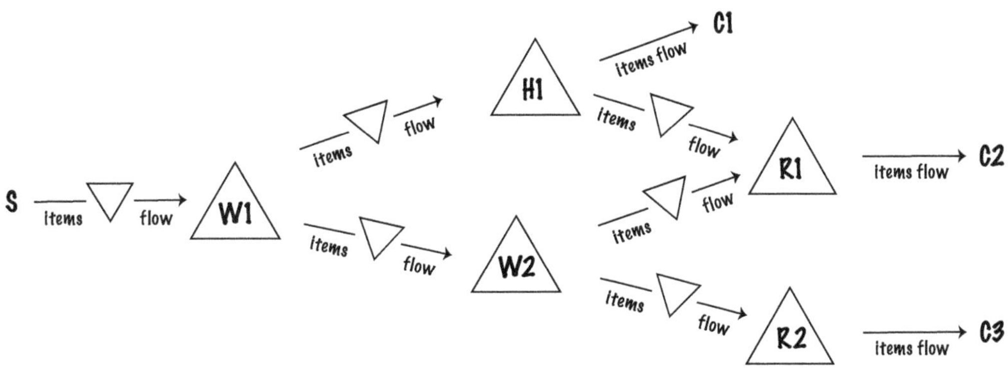

Fig. 119 A general multi-echelon system, in which installations can have multiple successors as well as predecessors.

Due to their very general structure, it is difficult to derive properties and/or insights on the nature and structure of the optimal control policy for such systems.

For instance, consider the following difficulty: in production it is common to assemble multiple components into a final or intermediate product. In the case of serial and assembly systems, this complication is only apparent. In these systems, a component can only be part of a given item. Therefore we can always redefine the unit of measure: e.g. if a given item contains two units of a given component, we can define two units of this component to be the new unit. However, this is not possible in general multi-echelon system such as the one in Fig. 119. To see this, consider installation *H*1, where *H* denotes a "hybrid" installation that can serve as a retailer but also as a warehouse. *H*1 can sell components directly to customers C1, or it may supply these components to *R*1. It may happen that the product manufactured by *R*1 contains two units of the component supplied by *H*1. In this case, it is clear it is not possible to avoid the difficulty by redefining the unit.

Appendix

https://doi.org/10.11647/OBP.0252.07

Introduction

In this Appendix, we provide relevant formal background on Poisson processes, Discrete Event Simulation, and Stochastic Dynamic Programming.

Topics

Poisson processes

IN THIS SECTION we discuss the nature and properties of Poisson processes.

A typical example of an experiment that can be modelled via a Bernoulli random variable is the toss of a coin; assuming the coin is fair, then $p = 0.5$.

Definition 28. *A Bernoulli random variable takes value* 1 *with probability* p*, and* 0 *with probability* $q \triangleq p - 1$.

Definition 29. *A Bernoulli stochastic process is a collection* $\{X_1, X_2, \ldots\}$ *of Bernoulli random variables.*

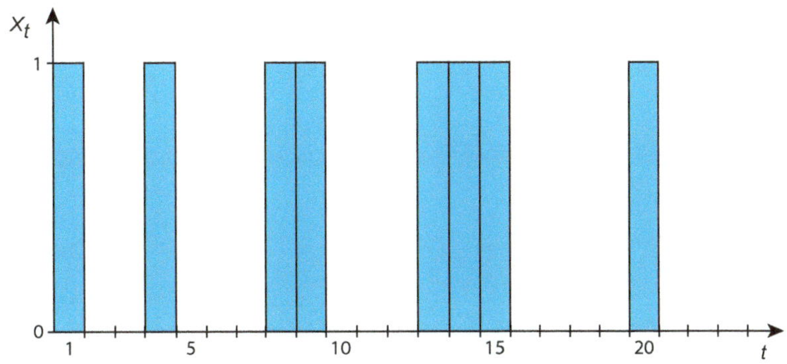

Fig. 120 A Bernoulli stochastic process.

In Fig. 120 we illustrate the dynamics of a Bernoulli process.

Definition 30. *A binomial random variable with parameters* n *and* p *is the sum of* n *Bernoulli random variables with parameters* p.

Let X be a binomial random variable, its probability mass function is then

$$\Pr(X = k) = \binom{n}{k} p^k q^{n-k};$$

its cumulative distribution function is

$$F(x) = \sum_{k=0}^{x} \binom{n}{k} p^k q^{n-k}.$$

Definition 31. *A binomial stochastic process is a collection* $\{X_1, X_2, \ldots\}$ *of Binomial random variables.*

In Fig. 121 we illustrate the dynamics of a binomial process.

In inventory control, consider a series of time periods and a setting in which in each time period it is possible to observe at most one unit of demand with probability p. A Bernoulli random variable models the occurrence of one unit of demand in any given period. A binomial random variable models the total demand observed in a sequence of n independent periods.

Definition 32. *A Poisson random variable with parameters* λ *is the limiting case of a binomial random variable when* $n = \infty$ *and* $np = \lambda$.

Let X be a Poisson random variable, its probability mass function is then

$$\Pr(X = k) = \frac{\lambda^k e^{-\lambda}}{k!};$$

its cumulative distribution function is

$$F(x) = \sum_{k=0}^{x} \frac{\lambda^k e^{-\lambda}}{k!}.$$

A Poisson random variable therefore models a system with a large number of possible events, each of which is rare; events occur with a known constant mean rate λ and independently of the time since the last event occurred.

Unfortunately, it is difficult to interpret a Poisson process in a way similar to that used for a binomial process. In fact, units of demand may occur at arbitrary positive times, and the probability p of a unit of demand at any particular instant is infinitely small. This means that there is no very clean way of describing a Poisson process in terms of the probability of an arrival at any given instant.

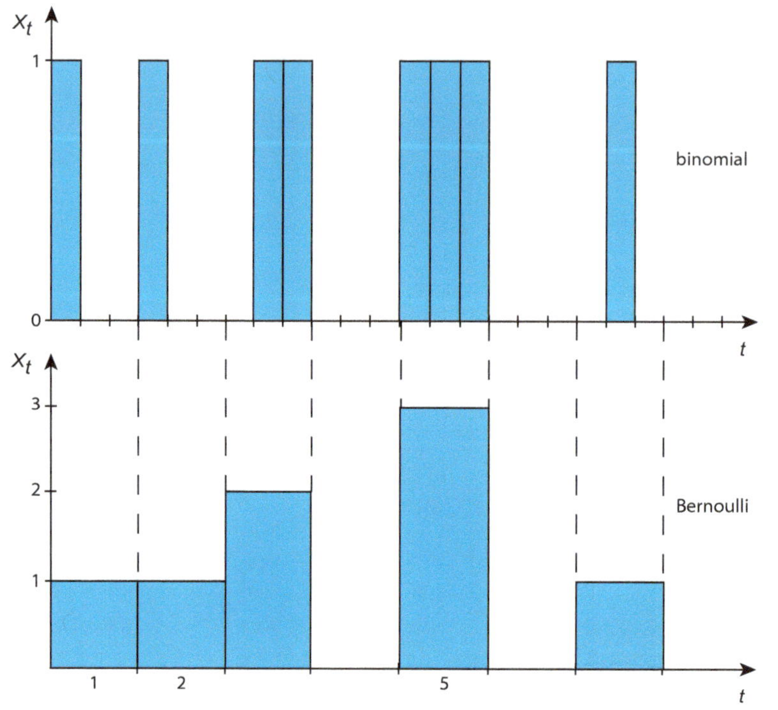

Fig. 121 A binomial stochastic process where $n = 3$, and its underpinning Bernoulli stochastic process.

It is more convenient to define a Poisson process variable in terms of the sequence of interarrival times, $X_1, X_2, \ldots$, which are defined to be independently and identically distributed random variables.

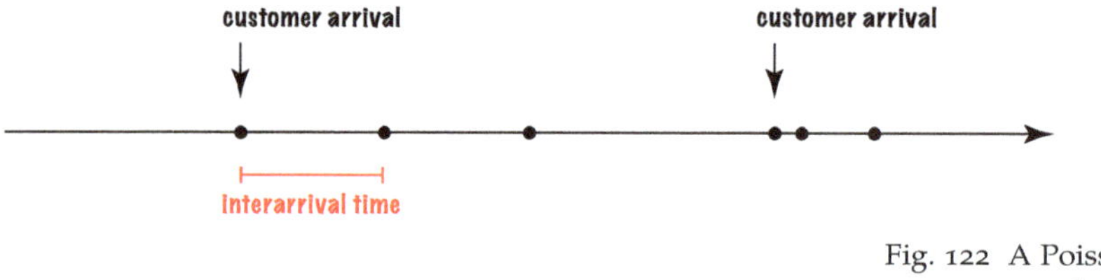

Fig. 122 A Poisson stochastic process seen as an arrival process in terms of interarrival times between successive events.

Definition 33. *An arrival process $\{S_1, S_2, \ldots\}$ is a sequence of increasing random variables, that is $0 < S_1 < S_2 < \ldots$, which are called arrival epochs. In this process, random variable $X_i \triangleq S_{i+1} - S_i$, which is called the interarrival time between event i and event $i+1$, is a positive random variable, that is $Pr(X_i \leq 0) = 0$.*

Condition $\Pr(X_i \leq 0) = 0$ effectively means that events cannot occur simultaneously.

Definition 34. *A renewal process is an arrival process for which the sequence of interarrival times is a sequence of independent and identically distributed random variables.*

Definition 35. *A Poisson process is a renewal process in which the interarrival intervals follow an exponential distribution; i.e. for some real $\lambda > 0$, each X_i has the probability density function $f(x) = \lambda e^{-\lambda x}$.*

Let X be an exponential random variable, its probability density function is

$$f(x) = \lambda e^{-\lambda x};$$

its cumulative distribution function is

$$F(x) = 1 - e^{\lambda x}.$$

The parameter λ is called the rate of the process; for any interval of size t, λt is the expected number of arrivals in that interval. Thus λ is called the arrival rate of the process.

Definition 36. *A random variable X possesses the memoryless property if it is positive, i.e. $Pr(X \leq 0) = 0$, and for every $s \geq 0$ and $t \geq 0$*

$$Pr(X \geq s + t | X > s) = Pr(X \geq t). \tag{44}$$

Lemma 58. *A continuous (resp. discrete) random variable X is exponential (resp. geometric) if and only if it possesses the memoryless property.*

Proof. This proof will only cover the continuous case.

To show that ($\rightarrow$) if X is exponential, then it possesses the memoryless property, observe that

$$\begin{aligned}
\Pr(X \geq s + t | X > s) &= \frac{\Pr(X \geq s + t \cap X > s)}{\Pr(X > s)} \\
&= \frac{\Pr(X \geq s + t)}{\Pr(X > s)} \\
&= \frac{e^{-\lambda(s+t)}}{e^{-\lambda s}} \\
&= e^{-\lambda t} \\
&= \Pr(X \geq t)
\end{aligned}$$

We next show that ($\leftarrow$) if X possesses the memoryless property, then it is exponential. Observe that $\Pr(X \geq s + t | X > s)\Pr(X \geq s) = \Pr(X \geq s + t)$, then

$$\Pr(X \geq s + t) = \Pr(X \geq t)\Pr(X \geq s). \tag{45}$$

Let $h(x) = \ln(\Pr(X \geq x))$ and observe that since $\Pr(X \geq x)$ is nonincreasing in x, $h(x)$ is also. Moreover, Eq. 45 implies that $h(s + t) = h(s) + h(t)$ for all $s \geq 0$ and $t \geq 0$. These two statements imply that $h(x)$ must be linear in x, and $\Pr(X \geq x)$ must be exponential in x. □

Let X be the waiting time until some given arrival, then Eq. 44 states that, given that the arrival has not occurred by time t, the distribution of the remaining waiting time (given by x on the left side of Eq. 44) is the same as the original waiting time distribution (given on the right side of Eq. 44), i.e. the remaining waiting time has no "memory" of previous waiting.

From Definition 35 and Lemma 58, it immediately follows that the portion of a Poisson process starting at an arbitrary time $t > 0$ is a probabilistic replica of the process starting at 0; that is, the time until the first arrival after t is an exponentially distributed random variable with parameter λ, all subsequent arrivals are independent of this first arrival and of each other, and all have the same exponential distribution.

Lemma 59. *Let X and Y be independent Poisson random variables with rates λ_X and λ_Y, respectively; then $X + Y$ is a Poisson random variable with rate $\lambda_X + \lambda_Y$.*

Proof. Let $Z = X + Y$ and $\lambda = \lambda_X + \lambda_Y$, then $\Pr(Z = z) = \sum_{j=0}^{z} \Pr(X = j)\Pr(Y = z - j) = \frac{e^{-\lambda}}{z!} \sum_{j=0}^{z} \binom{z}{j} \lambda_X^j \lambda_Y^{z-j} = \frac{e^{-\lambda}}{z!} (\lambda_X + \lambda_Y)^z$. The last step, which was obtained by using binomial expansion, concludes the proof: $\Pr(Z = z) = \frac{e^{-\lambda}}{z!} \lambda^z$. □

Discrete Event Simulation

A QUEUEING SYSTEM is a generic model that captures a variety of real-world scenarios: ticket offices, call centers, etc.

We consider a single teller who provides a service requiring a certain **service time** to be completed; customers arrive randomly and, if the teller is already busy, wait in a queue. The dynamics of the system are shown in Fig. 123.

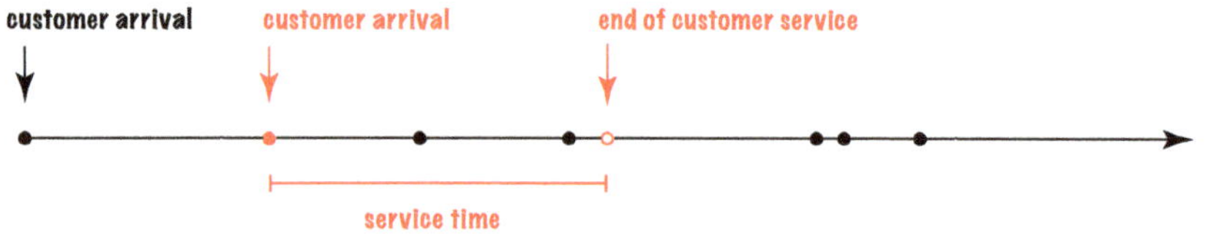

Fig. 123 The dynamics of the queueing system.

```
import matplotlib.pyplot as plt
import numpy as np
import pandas as pd

def plot_queue(values, label):

    # data
    df=pd.DataFrame({'x':
           np.array(values)[:,0], 'fx':
           np.array(values)[:,1]})

    # plot
    plt.xticks(range(len(values)),
              range(1,len(values)+1))
    plt.ylim(min(np.array(values)[:,1]),
           max(np.array(values)[:,1]))
    plt.xlabel("t")
    plt.ylabel("customers")
    plt.plot( 'x', 'fx', data=df,
           linestyle='-', marker='o',
           label=label)
```

Listing 75 Plotting the queue length in Python.

THE QUEUE displays the behaviour shown in Fig. 124.

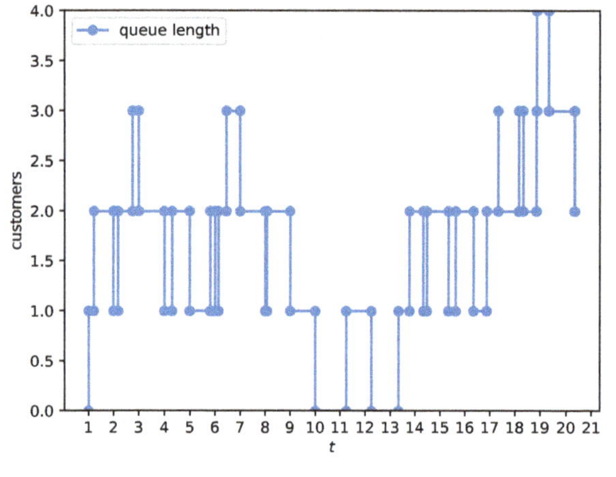

Fig. 124 The behaviour of the queue simulated in our numerical example.

DISCRETE EVENT SIMULATION (DES) is a modelling framework that can be used to model queueing systems. The DES model for the queueing system previously described is captured by the flow diagram[69] in Fig. 125.

[69] Arnold H. Buss. A tutorial on discrete-event modeling with simulation graphs. In C. Alexopoulos, I Kang, W. R. Lilegdon, and D. Goldsman, editors, *Proceedings of the 1995 Winter Simulation Conference ed.*, Arlington, Virginia, 1995.

THE QUEUE is the scarce resource modelled by class `Queue`: it features a `push` method and a `pop` method to add and remove customers. It also features a `review_queue` method to record the length of the queue at a given point in time.

THE DES ENGINE is captured in class `DES`. The engine comprises a method `start` that initates the loop by which the engine extracts events from the event stack `events`, and executes method `end` of for each of them. Finally, the engine comprises a method `schedule` to schedule an event after a given time lag.

Listing 76 A DES to model a queueing system in Python.

```
from queue import PriorityQueue

##############################
##    QUEUEING SYSTEM   ##
##############################

class Queue:
   def __init__(self, initial_size):
      self.s = 0
      self.levels = [[0, self.s]]

   def push(self):
      self.s += 1

   def pop(self):
      self.s -= 1

   def queue_empty(self):
      return self.s == 0

   def review_queue(self, time):
      self.levels.append([time, self.s]) # review queue

##########################
##      DES ENGINE      ##
##########################

class Event():
   pass

class EventWrapper():
   def __init__(self, event):
      self.event = event

   def __lt__(self, other):
      return self.event.priority < other.event.priority # low numbers denote high
           priority

class DES():
   def __init__(self, end):
      self.events, self.end, self.time = PriorityQueue() , end, 0

   def start(self):
      while True: # cycle until self.time < self.end
         event = self.events.get() # extract an event
         self.time = event[0]
         if self.time < self.end:
            event[1].event.end() # call method end() for the event
         else:
            break

   def schedule(self, event: EventWrapper, time_lag): # schedule an event
      self.events.put((self.time + time_lag, event))

###########################
##       EVENTS       ##
###########################

class CustomerArrival(Event):
   def __init__(self, des: DES, service_time: float, arrival_rate: float, queue:
        Queue):
      self.des, self.t, self.r, self.q = des, service_time, arrival_rate, queue
      self.priority = 1 # lowest priority

   def end(self):
      self.q.review_queue(self.des.time)
      if self.q.queue_empty():
         self.des.schedule(EventWrapper(EndOfCustomerService(self.des, self.t,
              self.q)), self.t)
      self.q.push()
      self.q.review_queue(self.des.time)
      self.des.schedule(EventWrapper(self), np.random.exponential(1.0/self.r)) #
           schedule another arrival

class EndOfCustomerService(Event):
   def __init__(self, des: DES, service_time: float, queue: Queue):
      self.des, self.t, self.q = des, service_time, queue
      self.priority = 0 # highest priority

   def end(self):
      self.q.review_queue(self.des.time)
      self.q.pop()
      if not(self.q.queue_empty()):
         # schedule end of customer service after self.t time periods
         self.des.schedule(EventWrapper(EndOfCustomerService(self.des, self.t,
              self.q)), self.t)
      self.q.review_queue(self.des.time)
```

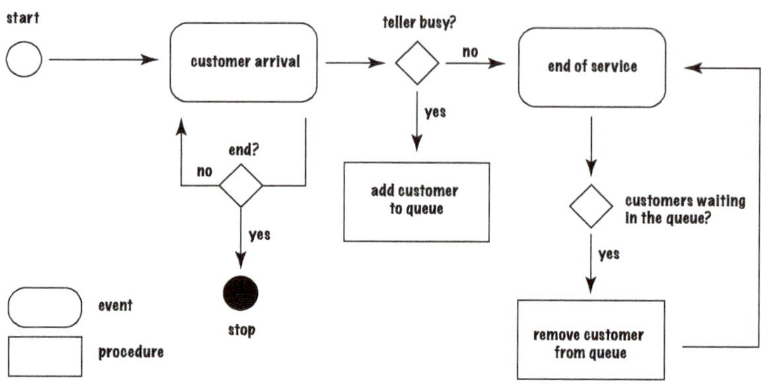

Fig. 125 DES flow diagram for the queueing system.

EVENTS are necessary to properly capture the dynamics of the system. `Event` is an empty class capturing a generic event, `EventWrapper` is a wrapper class that allows us to express relative priorities between event types. For instance, suppose that in an inventory system with no stock available, an order receipt event is scheduled at the same time as a customer demand, which event should be processed first? The field `priority` in each event class allows us to break such a tie.

THERE ARE TWO MAIN EVENTS we need to consider: `CustomerArrival` and `EndOfCustomerService`. `CustomerArrival` models the random arrival of a customer. The `end` method schedules an `EndOfCustomerService` event if the queue is empty, otherwise it increments the number of customers in the queue. Finally, it triggers a new `CustomerArrival` event after a time interval exponentially distributed with rate parameter equal to `customer_arrival_rate`. The `end` method of `EndOfCustomerService` event, decrements the number of customers in the queue, and if the queue is not empty, triggers another `EndOfCustomerService` event after `customer_service_time` time periods.

THE PYTHON CODE for simulating our example is shown in Listings 75, 76, and 77. After running the code, we find that the mean queue length is 1.75 customers (standard deviation: 0.93) and that the maximum observed queue length over the simulation horizon is 4 customers.

Listing 77 Simulating a queueing system in Python.

```
np.random.seed(1234) # set a random seed to ensure replicability
q = Queue(0) # create an empty queue
N = 20 # simulation horizon
des = DES(N)
customer_service_time, customer_arrival_rate = 1, 1
d = CustomerArrival(des, customer_service_time, customer_arrival_rate, q)
des.schedule(EventWrapper(d), 0) # schedule the first arrival
des.start()

print("Mean queue length:\t"+ '%.2f' % np.average(np.array(q.levels)[:,1]))
print("St. dev. queue length:\t"+ '%.2f' % np.std(np.array(q.levels)[:,1]))
print("Max queue length:\t"+ '%.2f' % max(np.array(q.levels)[:,1]))

plot_queue(q.levels, "queue length")
plt.legend()
plt.show()
```

Stochastic Dynamic Programming

DYNAMIC PROGRAMMING is a framework for modeling and solving sequential decision making problems. The framework was originally introduced by Bellman in his seminal book *Dynamic Programming*[70] to deal with multistage decision processes under uncertainty. The framework takes a "functional equation" approach to the discovery of optimum policies. Although originally devised to deal with problems of decision making under uncertainty, dynamic programming can also solve deterministic problems.

[70] Richard Bellman. *Dynamic Programming*. Princeton Univ. Pr., 1957.

TO MODEL AND SOLVE a problem via dynamic programming, one has to specify:

- a **planning horizon** comprising n periods;
- the finite set S_t of possible **states** in which the system may be found in period t, for $t = 1, ..., n$;
- the finite set A_s of possible **actions** that may be taken in state $s \in S_t$;
- the **state transition function** $g_t : S_t \times A_s \rightarrow S_{t+1}$ that identifies the state $s' \in S_{t+1}$ towards which the system transitions if action $a \in A_s$ is taken in state $s \in S_t$;
- the **immediate cost** (resp. profit) $c_t(s, a)$ incurred if action $a \in A_s$ is taken in state $s \in S_t$ of period t;
- the **functional equation** $f_t(s)$ denoting the minimum total cost (resp. maximum total profit) incurred over periods $t, t+1, \ldots, T$ when the system is in state $s \in S_t$ at the beginning of period t.

Without loss of generality in what follow we will consider a cost minimisation setting.

IN DETERMINISTIC DYNAMIC PROGRAMMING one usually deals with functional equations taking the following structure

$$f_t(s) = \min_{a \in A_s} c_t(s, a) + f_{t+1}(g_t(s, a)),$$

where the boundary condition of the system is $f_{T+1}(s) \triangleq 0$, for all $s \in S_{T+1}$. Let the initial state of the system at the beginning of the first period be s, the goal is to determine $f_n(s)$.

Given the current state s and the current action a in period t, we know with certainty the cost during the current stage and — thanks to the state transition function g_t — the future state towards which the system transitions.

In practice, however, even if we know the state of the system at the beginning of the current stage as well as the decision taken, the state of the system at the beginning of the next stage and the current period reward are often random variables that can only be observed at the end of the current stage.

STOCHASTIC DYNAMIC PROGRAMMING deals with problems in which the current period reward and/or the next period state are random, i.e. with multi-stage stochastic systems. The decision maker's goal is to maximise expected (discounted) reward over a given planning horizon. In their most general form, stochastic dynamic programs deal with functional equations taking the following structure

$$f_t(s) = \min_{a \in A_s} c_t(s,a) + \alpha \sum_{j \in S_{t+1}} p^a_{sj} f_{t+1}(j).$$

where

- $c_t(s,a)$ is the **expected immediate cost** (resp. profit) incurred if action $a \in A_s$ is taken in state $s \in S_t$ of period t;
- α is the **discount factor**;
- p^a_{sj} be the **transition probability** from state $s \in S_t$ towards state $j \in S_{t+1}$, when action $a \in A_i$ is taken;
- $f_t(s)$ is the minimum **expected total cost** (resp. maximum total profit) that can be attained during stages $t, t+1, \ldots, n$, if the system is in state s at the beginning of period t.

Let the initial state of the system at the beginning of the first period be s, once more the goal is to determine $f_n(s)$.

Example 40. *Consider a 3-period inventory control problem. At the beginning of each period the firm should decide how many units of a product should be produced. If production takes place for x units, where $x > 0$, we incur a production cost $c(x)$. This cost comprises both a fix and a variable component: $c(x) = 0$, if $x = 0$; $c(x) = 3 + 2x$, otherwise. Production in each period cannot exceed 4 units. Demand in each period takes two possible values: 1 or 2 units with equal probability (0.5). Demand is observed in each period only after production has occurred. After meeting current period's demand holding cost of \$1 per unit is incurred for any item that is carried over from one period to the next. Because of limited capacity the inventory at the end of each period cannot exceed 3 units. All demand should be met on time (no backorders). If at the end of the planning horizon (i.e. period 3) the firm still has units in stock, these can be salvaged at \$2 per unit. The initial inventory is 1 unit.*

The problem described in the previous example can be implemented in Python as shown in the `InventoryControl` class below. Additional classes used are shown in Listing 78 and Listing 79. The following code captures the instance described.

```
instance = {"T": 3, "K": 3, "v": 2, "h": 1, "s": 2, "pmf": [[(1, 0.5),(2, 0.5)] for
    i in range(0,3)], "C": 3}
ls = InventoryControl(**instance)
t = 0    # initial period
i = 1    # initial inventory level
print("f_1("+str(i)+"): " + str(ls.f(i)))
print("b_1("+str(i)+"): " + str(ls.q(t, i)))
```

```
import functools

class memoize(object):

    def __init__(self, func):
        self.func = func
        self.memoized = {}
        self.method_cache = {}

    def __call__(self, *args):
        return self.cache_get(
            self.memoized, args,
            lambda: self.func(*args))

    def __get__(self, obj, objtype):
        return self.cache_get(
            self.method_cache, obj,
            lambda: self.__class__(
                functools.partial(
                    self.func, obj)))

    def cache_get(self, cache, key,
            func):
        try:
            return cache[key]
        except KeyError:
            cache[key] = func()
            return cache[key]

    def reset(self):
        self.memoized = {}
        self.method_cache = {}
```

Listing 78 The `Memoize` class; **memoization** is a technique for storing the results of expensive function calls and returning the cached result when the same inputs occur again.

```
class State:
    '''the state of the inventory system
    '''

    def __init__(self, t: int, I:
            float):
        '''state constructor

        Arguments:
            t {int} -- time period
            I {float} -- initial inventory
        '''
        self.t, self.I = t, I

    def __eq__(self, other):
        return self.__dict__ ==
            other.__dict__

    def __str__(self):
        return str(self.t) + " " +
            str(self.I)

    def __hash__(self):
        return hash(str(self))
```

Listing 79 `State` class.

```
class InventoryControl:
   '''the inventory control problem

   Returns:
      [type] -- [description]
   '''

   def __init__(self, T:int, K: float, v: float, h: float, s: float, pmf:
        List[List[Tuple[int, float]]], C: float):
      '''inventory control problem constructor

      Arguments:
         T {int} -- periods in planning horizon
         K {float} -- fixed ordering cost
         v {float} -- per item ordering cost
         h {float} -- per item holding cost
         s {float} -- per item salvage value
         pmf {List[List[Tuple[int, float]]]} -- probability mass function
         C {float} -- capacity of the warehouse
      '''

      self.max_demand = max([max(i, key=lambda x: x[0])[0] for i in pmf])
      self.min_demand = min([min(i, key=lambda x: x[0])[0] for i in pmf])
      self.max_order_qty = C + self.min_demand

      # initialize instance variables
      self.T, self.K, self.v, self.h, self.s, self.pmf, self.warehouseCapacity = T,
           K, v, h, s, pmf, C

      # lambdas
      self.ag = lambda s: [i for i in range(max(self.max_demand - s.I, 0),
                           min(self.max_order_qty - s.I, self.max_order_qty)
                                + 1)] # action generator
      self.st = lambda s, a, d: State(s.t+1, s.I+a-d) # state transition
      L = lambda i,a,d : self.h*max(i+a-d, 0)         # immediate holding cost
      S = lambda i,a,d : self.s*max(i+a-d, 0)         # immediate salvage value
      self.iv = lambda s, a, d: ((self.K + self.v*a if a > 0 else 0) +
               L(s.I, a, d) -
               (S(s.I, a, d) if s.t == T - 1 else 0)) # immediate value function

      self.cache_actions = {}                         # cache with optimal
           state/action pairs

   def f(self, level: List[float]) -> float:
      s = State(0,level)
      return self._f(s)

   def q(self, period: int, level: List[float]) -> float:
      s = State(period,level)
      return self.cache_actions[str(s)]

   @memoize
   def _f(self, s: State) -> float:
      #Forward recursion
      v = min(
         [sum([p[1]*(self.iv(s, a, p[0])+                 # immediate cost
                 (self._f(self.st(s, a, p[0])) if s.t < self.T - 1 else 0)) #
                      future cost
             for p in self.pmf[s.t]])                     # demand realisations
          for a in self.ag(s)])                           # actions

      opt_a = lambda a: sum([p[1]*(self.iv(s, a, p[0])+
                      (self._f(self.st(s, a, p[0])) if s.t < self.T - 1 else 0))
                  for p in self.pmf[s.t]]) == v
      q = [k for k in filter(opt_a, self.ag(s))]          # retrieve best action
           list
      self.cache_actions[str(s)]=q[0] if bool(q) else None  # store an action in
           dictionary

      return v                                            # return expected total
           cost
```

To implement Stochastic Dynamic Programming in Python we make extensive use of `lambda` expressions (also known as Anonymous functions) to define: the function that generates the set of feasible actions for a given state, the state transition function that determines the future state given a present state and an action, and the immediate value function for an action taken in a given state.

MARKOV DECISION PROCESSES[71] represent a special class of stochastic dynamic programs in which the underlying stochastic process is a stationary process that features the Markov property.

[71] Martin L. Puterman. *Markov decision processes: Discrete stochastic dynamic programming*. J. Wiley & Sons, 1994.

Bibliography

Esther Arkin, Dev Joneja, and Robin Roundy. Computational complexity of uncapacitated multi-echelon production planning problems. *Operations Research Letters*, 8(2):61–66, 1989.

Kenneth J. Arrow. *Studies in the mathematical theory of inventory and production*. Stanford Univ. Pr., 1977.

Kenneth J. Arrow, Theodore Harris, and Jacob Marschak. Optimal inventory policy. *Econometrica*, 19(3):250–272, 1951.

Ronald G. Askin. A procedure for production lot sizing with probabilistic dynamic demand. *AIIE Transactions*, 13(2):132–137, 1981.

Sven Axsäter. Simple solution procedures for a class of two-echelon inventory problems. *Operations Research*, 38(1):64–69, 1990.

Richard Bellman. *Dynamic Programming*. Princeton Univ. Pr., 1957.

William A. Bernstein. Luca pacioli the father of accounting. In *The Air Force Comptroller*, volume 10(2) of *Air Force recurring publication 170-2*, pages 44–45. Office of the Comptroller, United States Air Force, 1976.

John Birge. *Introduction to stochastic programming*. Springer, 2011.

James H. Bookbinder and Jin-Yan Tan. Strategies for the probabilistic lot-sizing problem with service-level constraints. *Management Science*, 34(9):1096–1108, 1988.

George. E. P. Box and David R. Cox. An analysis of transformations. *Journal of the Royal Statistical Society: Series B (Methodological)*, 26(2): 211–243, 1964.

George. E. P. Box and Gwilym M. Jenkins. *Time series analysis: Forecasting and control*. Holden-Day, 1976.

Robert G. Brown. *Statistical forecasting for inventory control*. McGraw-Hill, 1959.

Arnold H. Buss. A tutorial on discrete-event modeling with simulation graphs. In C. Alexopoulos, I Kang, W. R. Lilegdon, and D. Goldsman, editors, *Proceedings of the 1995 Winter Simulation Conference ed.*, Arlington, Virginia, 1995.

Andrew J. Clark and Herbert Scarf. Optimal policies for a multi-echelon inventory problem. *Management Science*, 6(4):475–490, 1960.

Alfred Crosby. *The measure of reality: quantification and Western society, 1250-1600*. Cambridge Univ. Pr., 1997.

Edsger W. Dijkstra. A note on two problems in connexion with graphs. *Numerische Mathematik*, 1(1):269–271, 1959.

Gozdem Dural-Selcuk, Roberto Rossi, Onur A. Kilic, and S. Armagan Tarim. The benefit of receding horizon control: Near-optimal policies for stochastic inventory control. *Omega*, 97:102091, 2020.

James Durbin and Siem Jan Koopman. *Time series analysis by state space methods*. Oxford Univ. Pr., 2001.

Francis Y. Edgeworth. The mathematical theory of banking. *Journal of the Royal Statistical Society*, 51(1):113–127, 1888.

Donald Erlenkotter. Ford Whitman Harris and the Economic Order Quantity model. *Operations Research*, 38(6):937–946, 1990.

Awi Federgruen and Paul Zipkin. An inventory model with limited production capacity and uncertain demands I. The average-cost criterion. *Mathematics of Operations Research*, 11(2):193–207, 1986.

Michael Florian and Morton Klein. Deterministic production planning with concave costs and capacity constraints. *Management Science*, 18(1):12–20, 1971.

Michael Florian, Jan K. Lenstra, and Alexander H. G. Rinnooy Kan. Deterministic production planning: Algorithms and complexity. *Management Science*, 26(7):669–679, 1980.

Robert G. Gallager. *Stochastic processes*. Cambridge Univ. Pr., 2013.

Ford W. Harris. How many parts to make at once. *Factory, The Magazine of Management*, 10(2):135–136, 1913.

Charles C. Holt. Forecasting seasonals and trends by exponentially weighted moving averages. *International Journal of Forecasting*, 20(1): 5–10, 2004.

Rob J. Hyndman and George Athanasopoulos. *Forecasting: Principles and practice*. OTexts, 2020.

Peter Jackson, William Maxwell, and John Muckstadt. The joint replenishment problem with a powers-of-two restriction. *IIE Transactions*, 17(1):25–32, 1985.

Hamed Jalali and Inneke Van Nieuwenhuyse. Simulation optimization in inventory replenishment: A classification. *IIE Transactions*, 47(11):1217–1235, 2015.

Herman Kahn and Andy W. Marshall. Methods of reducing sample size in Monte Carlo computations. *Journal of the Operations Research Society of America*, 1(5):263–278, 1953.

William Karush. A theorem in convex programming. *Naval Research Logistics Quarterly*, 6(3):245–260, 1959.

Retsef Levi, Georgia Perakis, and Joline Uichanco. The data-driven newsvendor problem: New bounds and insights. *Operations Research*, 63(6):1294–1306, 2015.

Xiyuan Ma, Roberto Rossi, and Thomas Welsh Archibald. MILP approximations for non-stationary stochastic lot-sizing under (s,Q)-type policy, 2020. URL `https://arxiv.org/abs/2009.06976`.

Douglas C. Montgomery and George C. Runger. *Applied statistics and probability for engineers*. John Wiley and Sons, 2014.

John A. Nelder and Roger Mead. A simplex method for function minimization. *The Computer Journal*, 7(4):308–313, 1965.

Martin L. Puterman. *Markov decision processes: Discrete stochastic dynamic programming*. J. Wiley & Sons, 1994.

Paolo Quattrone. Books to be practiced: Memory, the power of the visual, and the success of accounting. *Accounting, Organizations and Society*, 34(1):85–118, 2009.

Paolo Quattrone. Governing social orders, unfolding rationality, and Jesuit accounting practices. *Administrative Science Quarterly*, 60 (3):411–445, 2015.

Jack Rogers. A computational approach to the economic lot scheduling problem. *Management Science*, 4(3):264–291, 1958.

Kaj Rosling. Optimal inventory policies for assembly systems under random demands. *Operations Research*, 37(4):565–579, 1989.

Francesca Rossi, Peter van Beek, and Toby Walsh, editors. *Handbook of Constraint Programming*, volume 2 of *Foundations of Artificial Intelligence*. Elsevier, 2006.

Roberto Rossi, Steven Prestwich, S. Armagan Tarim, and Brahim Hnich. Confidence-based optimisation for the newsvendor problem under binomial, Poisson and exponential demand. *European Journal of Operational Research*, 239(3):674–684, 2014a.

Roberto Rossi, S. Armagan Tarim, Steven Prestwich, and Brahim Hnich. Piecewise linear lower and upper bounds for the standard normal first order loss function. *Applied Mathematics and Computation*, 231:489–502, 2014b.

Roberto Rossi, Onur A. Kilic, and S. Armagan Tarim. Piecewise linear approximations for the static–dynamic uncertainty strategy in stochastic lot-sizing. *Omega*, 50:126–140, 2015.

Herbert E. Scarf. Optimality of (s, S) policies in the dynamic inventory problem. In K. J. Arrow, S. Karlin, and P. Suppes, editors, *Mathematical Methods in the Social Sciences*, pages 196–202. Stanford Univ. Pr., 1960.

Chen Shaoxiang. The infinite horizon periodic review problem with setup costs and capacity constraints: A partial characterization of the optimal policy. *Operations Research*, 52(3):409–421, 2004.

Chen Shaoxiang and Marc Lambrecht. X-Y band and modified (s, S) policy. *Operations Research*, 44(6):1013–1019, 1996.

Craig C. Sherbrooke. Metric: A multi-echelon technique for recoverable item control. *Operations Research*, 16(1):122–141, 1968.

E. W. Taft. The most economical production lot. *The Iron Age*, 101: 1410–1412, 1918.

S. Armagan Tarim and Brian G. Kingsman. Modelling and computing (R_n, S_n) policies for inventory systems with non-stationary stochastic demand. *European Journal of Operational Research*, 174(1): 581–599, 2006.

Huseyin Tunc, Onur A. Kilic, S. Armagan Tarim, and Burak Eksioglu. A simple approach for assessing the cost of system nervousness. *International Journal of Production Economics*, 141(2): 619–625, 2013.

Huseyin Tunc, Onur A. Kilic, S. Armagan Tarim, and Roberto Rossi. An extended mixed-integer programming formulation and dynamic cut generation approach for the stochastic lot-sizing problem. *INFORMS Journal on Computing*, 30(3):492–506, 2018.

Tim J. van Kampen, Renzo Akkerman, and Dirk Pieter van Donk. SKU classification: A literature review and conceptual framework. *International Journal of Operations & Production Management*, 32(7): 850–876, 2012.

Arthur F. Veinott and Harvey M. Wagner. Computing optimal (s, S) inventory policies. *Management Science*, 11(5):525–552, 1965.

Andrea Visentin. *Computing policy parameters for stochastic inventory control using stochastic dynamic programming approaches*. PhD thesis, University College Cork, 2020.

Andrea Visentin, Steven Prestwich, Roberto Rossi, and S. Armagan Tarim. Computing optimal (R, s, S) policy parameters by a hybrid of branch-and-bound and stochastic dynamic programming. *European Journal of Operational Research*, 2021.

Harvey M. Wagner and Thomson M. Whitin. Dynamic version of the economic lot size model. *Management Science*, 5(1):89–96, 1958.

Jacob Wijngaard. An inventory problem with constrained order capacity. Technical report, TU Eindhoven, the Netherlands, 1972.

Peter R. Winters. Forecasting sales by exponentially weighted moving averages. *Management Science*, 6(3):324–342, 1960.

Mengyuan Xiang, Roberto Rossi, Belen Martin-Barragan, and S. Armagan Tarim. Computing non-stationary (s, S) policies using mixed integer linear programming. *European Journal of Operational Research*, 271(2):490–500, 2018.

Yu-Sheng Zheng and Awi Federgruen. Finding optimal (s, S) policies is about as simple as evaluating a single policy. *Operations Research*, 39(4):654–665, 1991.

Index

www.ingramcontent.com/pod-product-compliance
Ingram Content Group UK Ltd.
Pitfield, Milton Keynes, MK11 3LW, UK
UKHW061848041026
15591UKWH00015B/980

9 781800 641761